The Wireless Application Protocol (WAP)

A Wiley Tech Brief

Steve Mann
Scott Sbihli

Wiley Computer Publishing

John Wiley & Sons, Inc.

NEW YORK • CHICHESTER • WEINHEIM • BRISBANE • SINGAPORE • TORONTO

Publisher: Robert Ipsen
Editor: Carol Long
Associate Editor: Margaret Hendrey
Managing Editor: Micheline Frederick
Text Design & Composition: Benchmark Productions, Inc.

Designations used by companies to distinguish their products are often claimed as trademarks. In all instances where John Wiley & Sons, Inc., is aware of a claim, the product names appear in initial capital or ALL CAPITAL LETTERS. Readers, however, should contact the appropriate companies for more complete information regarding trademarks and registration.

This book is printed on acid-free paper. ∞

Copyright © 2000 by Steve Mann and Scott Sbihli. All rights reserved.

Published by John Wiley & Sons, Inc.

Published simultaneously in Canada.

No part of this publication may be reproduced, stored in a retrieval system or transmitted in any form or by any means, electronic, mechanical, photocopying, recording, scanning or otherwise, except as permitted under Sections 107 or 108 of the 1976 United States Copyright Act, without either the prior written permission of the Publisher, or authorization through payment of the appropriate per-copy fee to the Copyright Clearance Center, 222 Rosewood Drive, Danvers, MA 01923, (978) 750-8400, fax (978) 750-4744. Requests to the Publisher for permission should be addressed to the Permissions Department, John Wiley & Sons, Inc., 605 Third Avenue, New York, NY 10158-0012, (212) 850-6011, fax: (212) 850-6008, e-mail: PERMREQ@WILEY.COM.

This publication is designed to provide accurate and authoritative information in regard to the subject matter covered. It is sold with the understanding that the publisher is not engaged in professional services. If professional advice or other expert assistance is required, the services of a competent professional person should be sought.

Library of Congress Cataloging-in-Publication Data:

Mann, Steve, 1950–
    The wireless application protocol (WAP) : a Wiley tech brief / Steve Mann, Scott Sbihli.
        p. cm. — (Wiley tech brief series)
    Includes bibliographical references and index.
    ISBN 0-471-39992-2 (pbk. : alk. paper)
    1. Computer network protocols. 2. Wireless communication systems. I. Sbihli, Scott.
II> Title. III. Series

TK5105.55 .M363 2000
004.6'2--dc21

Printed in the United States of America.

10 9 8 7 6 5 4 3 2 1

KINGSTON UPON HULL CITY LIBRARIES	
B3714225O	
Cypher	29.01.01
004.62	£21.50
	HCT 3/01

Wiley Tech Brief Series

Contents

	Acknowledgments	ix
	About the Authors	xi
	Introduction	xiii
Chapter 1	**Wireless Data Primer**	1
	Concepts	1
	Spectrum and Frequencies	2
	PCS	3
	Circuit Switched versus Packet Data Connections	3
	Analog versus Digital	4
	Transports and Protocols	5
	The ISO Network Model	6
	Wireless Technologies	8
	AMPS and European Analog Cellular	8
	TDMA	9
	CDMA	9
	GSM	10
	CDPD	11
	Voice/Data Networks	11
	Future Wireless Communications	12
Chapter 2	**A Brief History of WAP**	13
	Origins	13
	The WAP Forum	16
	Forum Members	17
	Hardware Providers	18

	Cellular Service Providers	18
	WAP Infrastructure Creators	18
	Software Developers	18
	Content/Service Providers	19
	Available WAP Services	19
Chapter 3	**Why WAP?**	**21**
	The Great Convergence	21
	The Internet	21
	Wireless Technology	23
	Computing Power	23
	Social and Economic Forces	24
	WAP Device Characteristics	25
	The Need For WAP	26
Chapter 4	**An Overview of WAP**	**29**
	WAP in Action	29
	Web Transaction Model	33
	WAP Transaction Model	35
	WAP Step-By-Step	36
	WAP Architecture	37
	WAP Application Environment (WAE)	38
	Wireless Session Protocol (WSP)	38
	Wireless Transaction Protocol (WTP)	39
	Wireless Transport Layer Security (WTLS)	39
	Wireless Datagram Protocol (WDP)	39
	Bearers	39
	A Closer Look at WAE	40
	Microbrowser	40
	WML	41
	WMLScript	43
	Wireless Telephony Application Interface (WTAI)	44
Chapter 5	**The WAP Application Environment**	**45**
	The Microbrowser	45
	WML	46
	Elements and Attributes	46
	Decks and Cards	47
	WML Features	49
	Content	49
	Tasks and Events	51
	Data Entry	55
	Input Alternatives	56
	WMLScript	57

Chapter 6	**WAP Client Software, Hardware, and Web Sites**	**59**
	OEM Microbrowsers	59
	UP.Browser	60
	Ericsson WAP Browser	60
	Mobile Explorer	60
	AU-System	61
	Consumer Microbrowsers	61
	WAPMan	62
	WinWAP	63
	4thpass KBrowser	64
	WAP Devices	64
	Nokia 6210/6250	65
	Nokia 7110	66
	Motorola	67
	Ericsson R320/R380/MC218	68
	MobileAccess T250	70
	NeoPoint 1000/1600	71
	Consumer WAP Sites	72
Chapter 7	**WAP Gateways**	**75**
	A Note on Terminology	75
	WAP Gateway Services	76
	Finding a Gateway	78
	Security	79
	Privacy	79
	Integrity	79
	Authentication	79
	Non-Repudiation	80
	WAP's Security	80
Chapter 8	**Some WAP Profiles**	**81**
	exo-net	81
	Business Background	81
	Features	82
	WAP Background	82
	Technology and Development	83
	Project Status	83
	MainFreight	84
	Business Background	84
	WAP Background	84
	Technology and Development	84
	Project Status	85
	Sky City Hotels	85
	Business Background	85

WAP Background	85
Technology and Development	86
Retrospect	86
A Consumer Profile	86
Services	87
Theory versus Practice	88
What WAP Does Well	89

Chapter 9 Implementing an Enterprise WAP Strategy **91**

Requirements	91
Architecture	92
Design/Implementation	93
Testing	96
Summary	96

Chapter 10 The Future of WAP **97**

Problems with WAP	97
Screen Size	97
Navigation	98
Data Entry	98
Latency	98
Duplicate Content	99
Solving These Problems	99
Screen Size	99
Navigation	100
Data Entry	100
Latency	101
The Next Generation	101

Appendix A Resources **103**

Appendix B WAP WML **109**

Glossary **191**

Index **201**

Acknowledgments

Steve Mann

I would like to thank Carol Long at John Wiley & Sons for the chance to do this project, horrific schedule and all.

I'd also like to thank Scott Sbihli, my co-author and former contributing editor at *Handheld Systems* magazine, who agreed to get involved in this project, regardless of the aforementioned horrific schedule. He delivered the goods (down to the wire sometimes, but hey, delivery is delivery), a co-author's single most important quality. He did it with grace and good humor, a co-author's second most important quality.

To Betty, my wife, who stayed calm and cheerful while I did this project. Finally, to my parents who made me do my homework and eat my green vegetables, and my sister Lisa who actually opened the cover of my WAP development book.

Scott Sbihli

I would like to thank, my co-author, Steve Mann for providing me my first breaks as both a magazine writer and a book author. He has provided insight and guidance to my writing as well as being a first-class friend.

Dave and Phil at the Holliday Group in New Zealand for their extensive input on the Profile and Enterprise chapters. Christina Berry at John Wiley & Sons for handling all of the busy (but crucial) work related to obtaining permission to reprint certain product pictures in this book.

My girlfriend, Jahnavi, for her support over the past 10 weeks. She patiently listened to me drone on about WAP as it consumed my life. My business partners, Jeff and Richard, for supporting the creation of this book even though my day job demanded more than 40 hours a week.

Finally, to Mike, Amy, Sarah, Julie, and Davis for taking the time to ask how the writing was going and not deserting me because I've had little time to see them.

About the Authors

Steve Mann is a software developer, consultant, analyst, and writer specializing in mobile computing since 1993. He is the former publisher and editor of *Handheld Systems* magazine, the author of *Programming Applications with the Wireless Application Protocol*, and co-author of *Advanced Palm Programming: Developing Real World Applications*, both from John Wiley & Sons. He holds B.S. and M.S. degrees in computer science and has more than 25 years experience in the computer industry. He can be reached at smann@cdpubs.com.

Scott Sbihli is President of Empyrean Design Works Inc., a full-service consulting and development company serving the enterprise in the mobile, handheld, and wireless markets. He has 15 years of experience in the computing industry and has been a writer or contributing editor to *Pen Computing*, *Handheld Systems*, and *Handheld Computing* magazines. He can be reached at slsbihli@empyreandw.com.

Introduction

Apparently, the whole world is crazy about wireless communications. There are more than 400 million cellular phone users in the world as of mid-2000. Market forecasters are predicting that there will be more than one billion by the end of the year 2002. They also claim that many of these users will buy new telephones, not keep their old ones, to get the latest and greatest features.

That's a lot of new gizmos.

The telephone manufacturers, the infrastructure providers, the governments auctioning off spectrum, everyone involved in the cellular phone industry thinks they're going to end up rich, thin, and ready to retire. Some will, most won't.

What about cellular phone *users*? What makes everyone, including us, think that the number of cell phone users will double or triple in the next few years? (In addition to the exciting prospect of increasing your odds of being in an automobile accident, that is?)

The secret is the Internet. People are becoming more mobile and they are becoming more connected. But what they *really* want is to be connected while mobile.

Enter WAP, the Wireless Application Protocol. It was designed to give wireless telephone users a way to access Internet-based content using their wireless telephones. WAP is a collection of technologies—programming languages, communications protocols, infrastructure architectures, specifications, and more—that does just that.

Market forecasters predict that anywhere from 30 to 80 percent of all new cellular telephones sold in the next year will be WAP-capable. That's a lot of WAP gizmos, which also have the wireless hardware and software people salivating. Unfortunately, in this type of marketing feeding frenzy, it's usually difficult to find out just what the heck people are talking about.

That's where *The Wireless Application Protocol: A Wiley Tech Brief* comes in. In this book, we try to provide you with a relatively non-technical introduction to WAP, and answer such questions as: What is WAP? Does it work? Can my company derive any benefit from using WAP? Where can I find more information about WAP products and services?

Road Map

This book contains 10 chapters. The first, "Wireless Data Primer", gets you up to speed on wireless data communications basics—concepts, terminology, systems, acronyms, and the like. We use the content introduced in Chapter One throughout the rest of this book.

We strongly encourage you to start with this chapter. If you work in the wireless communications industry, you'll probably find Chapter One an easy, but hopefully informative, read.

Although we encourage you to proceed sequentially, the remainder of the chapters in this book can be read in any order.

Chapters Two and Three, "A Brief History of WAP" and "Why WAP?", give you some background on the birth of WAP and the WAP Forum, the international organization responsible for fostering WAP. We also discuss why many people feel that WAP is the direction in which wireless data is evolving.

Chapter Four, "An Overview of WAP", covers two topics. First, we show you what a WAP telephone actually looks like and how you might use it. We try to give you a feel for the user experience. After all, WAP is supposed to be about users. Second, we briefly describe WAP's major components at a high level. This information is digestible by anyone. You don't need to be a software engineer to understand it.

Chapter Five, "The WAP Application Environment", contains a more detailed look at WAP's application-development technologies. It also has a few short examples of WML, the primary WAP language. If you're interested in creating WAP applications, this is the chapter to read. Although not terribly technical, this chapter is best suited for people that have, for example, modest experience designing web pages or at least a minimal understanding of HTML.

We assume that in this day and age, everyone and their household pet has a web site, and a cursory knowledge of how web sites work. If you don't fit into that category, try browsing Chapter Five anyway. You might be pleasantly surprised. If you find yourself quickly drifting off, skip to another chapter. This chapter isn't essential to understanding how WAP works, its benefits, or its problems.

Chapter Six, "WAP Client Software, Hardware, and Web Sites", is a compendium of hardware, software, and web sites. We don't try to include all WAP-compatible products and sites. That would be too difficult, too much information, and obsolete before it ever arrived in a bookstore. Instead, we try to give you a flavor for the types of client products that are currently available plus web site addresses so you can find more on your own.

Chapter Seven, "WAP Gateways", discusses WAP gateways, Internet-based servers through which all WAP transactions must pass. These gateways are the hidden plumbing that makes WAP a reality. If you are an enterprise IT manager this chapter is essential reading.

Chapter Eight, "Some WAP Profiles", describes three enterprise WAP projects and one person's experience with a consumer WAP service. For the enterprise profiles, we focus on the business value and implementation issues. The consumer profile describes the services and the experience.

Chapter Nine, "Implementing an Enterprise WAP Strategy", complements Chapter Eight by discussing key issues if you're thinking of incorporating WAP into your company's IT portfolio.

Finally, Chapter Ten, "The Future of WAP", describes in general terms where we think the WAP market is going. We don't delve into all the gory details of future WAP specifications and the like. Instead, we focus on the big picture—hardware, software, and use patterns. We also discuss some of WAP's current problems.

We finish with an Appendix of resources. About 25 percent of them are printed or web-based documents. The rest are web sites—places to find more WAP information, pointers to products, services, and service providers, and the like. Like all web collections, these resources will probably change quite a bit by the time you read this book. Treat our list as a starting point.

Finally, we include an extensive Glossary of terms that encompass the cellular telephone industry, WAP, and the Internet. WAP is really about the convergence of wireless communications and the Internet. We wanted to provide you with an extensive glossary that covers all bases.

Who Should Read This Book

The most important audience for this book, the one that actually prompted its creation, includes people in various aspects of the wireless communications industry. These people have been hearing about this thing called WAP for months now and want some details. They need to be informed about new industry developments, but don't want a hard-core technical introduction to the subject. Instead, they need a quick introduction so they can intelligently discuss WAP's components, strengths, and weaknesses. They also need pointers to additional resources for more information.

With that audience in mind, we tried to make this book as general as possible. It's for anyone involved in any aspect of communications, the computer industry, or the Internet, or anyone who is curious about the technology firestorm that's raging (out of control some might say) around us. It's a primer, a getting started guide. If you use a computer or a cellular telephone, or even only know what they are, we think you'll find this book useful and informative.

CHAPTER 1

Wireless Data Primer

The Wireless Application Protocol (WAP) is designed to work on all major global *wireless communications systems*, the types of communications networks we describe in this chapter. Before we can discuss WAP in any detail, we need to cover some basic communications principles and describe the most popular of these systems.

First, a few caveats. This chapter isn't comprehensive—we only cover those concepts and systems you need to understand to make sense of WAP. It isn't thorough either—we only cover enough detail to make sense of the concepts and make them useful. In some cases, we may even fudge the truth a bit to facilitate explaining a concept.

This chapter won't make you an expert on wireless technologies. It should, however, make the avalanche of acronyms in the field of wireless communications seem more like a gentle snowfall.

Concepts

There are many different types of wireless communications systems in use today. Some are based on cellular telephone technologies. Others may use satellite or microwave technologies. In this section we cover some of the basic concepts you need to know in order to understand how WAP works.

Spectrum and Frequencies

The air around us is full of mostly invisible waves flying in all directions. These waves are different sizes, where size is a measure of the number of times a wave occurs in a second, also called *cycles per second*. A single *cycle* or *wave* is also called a *Hertz*, named for Heinrich Rudolph Hertz, the German physicist (1857–1894) who discovered this.

For instance, when you play your stereo, the sound waves normally range from about 20 Hertz (abbreviated Hz) to 20 Kilohertz (or thousands of Hertz, abbreviated kHz). You can also express these two numbers as 20 cycles per second and 20,000 cycles per second. They define the frequency range for sound.

The number of times a wave occurs per second is also called its frequency. Low-pitched sound waves, like those created when a jet takes off, have lower frequencies than high-pitched sound waves like the squeal of an automobile's brakes.

The full range of frequencies, from the very lowest (zero cycles per second) to the very highest (just less than an infinite number of cycles per second), is called the *electromagnetic spectrum*. It is divided into distinct regions. Audible sound is at the low end. After that come the radio frequencies, ranging from about 100 kHz to 100 GHz (Gigahertz, or one billion Hertz). Above that we find visible light, ultraviolet and X-ray waves, and then Gamma rays, which top out at 10^{25} Hertz.

All of the activity that we term *wireless data* uses a subset of the radio frequency spectrum from approximately 800 kHz to 2 GHz. Wireless data is usually identified as operating at some fixed frequency. In reality, it typically uses some range of frequencies. For instance, the AMPS transport, which we describe shortly, is usually described as operating at 800 MHz. In fact, it operates between 824 MHz and 890 MHz.

Radio frequencies don't travel great distances, usually dozens, or at the most, hundreds of miles. As a result, various governmental organizations representing individual countries and groups of countries have made decisions over the years as to who can use parts of the spectrum and how they can use it. This is one reason wireless communications is so complicated.

A classic example is the Federal Communications Commission (FCC) in the United States. They allocate wireless frequencies and dictate how they are used. As a result of their careful deliberations, most parts of the United States now have at least six different wireless communications carriers vying for the attention of potential customers. Additionally, the frequencies that have been allocated for various types of wireless communications do not match the frequencies allocated in other parts of the world, making global roaming much more complex than it should be.

In the upcoming sections, we describe various systems partly in terms of their frequency range. As you will see, the same system can exist in two different parts of the world on different frequencies, which can add to the confusion.

PCS

During the 1990s, the FCC in the United States allocated a new spectrum at 1900 MHz for use for so-called *Personal Communications Systems* (PCS), all-digital wireless communications systems. The FCC auctioned off this spectrum to a variety of large and small companies for developing and deploying new PCS systems. For instance, Sprint purchased some of this spectrum and is busy deploying the Sprint PCS wireless network.

Unfortunately, the term PCS has been widely misused and can be confusing. At its simplest level, it merely means a wireless communications networks operating at 1900 MHz.

For more information about PCS technology use in the United States, visit www.pcsdata.com.

Circuit Switched versus Packet Data Connections

In the 20th century, the telephone has been perhaps the most widely used means for communicating between individuals. We all know how telephones work. First, you decide to call someone and dial his or her telephone number. If the person you are calling is available and decides they want to talk to you (after caller ID or voice mail screening), they pick up the phone. The two of you talk for a while and then hang up when you're finished.

Although this is a very simple process, a telephone call has one key characteristic for our purposes: you and the person you're talking to tie up a telephone circuit. Once the call goes through, the telephone company assigns you your own wire that you get to use exclusively until you hang up.

In the old days, you actually did have your own wire. Thanks to modern computers and software that runs in the telephone company switching offices, you don't really get a single dedicated wire anymore in most cases. Regardless, as long as your call is active, you're using system capacity and making it unavailable for a different phone call. You control the circuit.

Compare a telephone conversation with a wireless ship-to-shore radio which, incidentally, usually operates at about 45 MHz. The circuit is always there and always active. It only gets used if the captain of a sailboat whose engine just died has to call for help. He turns on the transmitter, presses a button, and broadcasts a request for help. The broadcast ties up the circuit for a brief period of time, and then the circuit becomes available again for someone else's use (hopefully someone responding to the call for help).

These two situations demonstrate two basic types of communication: *circuit switched* and *packet data*. In the case of a telephone call, you control the circuit until you are finished. It is a persistent connection, it's all yours, and you can do whatever you want with it. When you hang up, the circuit gets switched to the next person who wants to make a call, hence the term circuit switched.

In the case of the floundering boat, the circuit is available at the same time to everyone who has a transmitter that works on that same frequency. If two people transmit messages simultaneously, they both get garbled. However, because the messages are short and intermittent, the chance of two messages colliding is minimal.

These short bursty messages are well-suited to packet data communications. With packet data, messages are short and consume a modest chunk of the channel called a packet. You only own the channel at the time you send the packet.

Although our ship-to-shore radio example describes wireless radio voice communication, packet data, as you might have suspected, is really used for transmitting data wirelessly. A packet is a single short message containing a modest amount of digital data. The actual format of the packet and amount of data in it depend on the packet data method being used.

You can generally categorize all wireless communications sessions as either circuit switched or packet data. Circuit switched wireless communication requires the use of a *wireless data modem*. Laptop computer users who want to dial up an Internet Service Provider (ISP) or a corporate server to wirelessly send and receive e-mail, commonly use this method. A circuit switched wireless session takes quite a bit of time to start and stop (also called *setup* and *teardown*).

A WAP device typically interacts with Internet servers in a set of short, compact exchanges separated by waiting periods of at least a few seconds. WAP is best suited to packet data connections, and the WAP protocols are optimized for that type of connection. Although there's no intrinsic reason why you can't use a circuit switched connection to access WAP servers, the time required to start and stop the connection, plus the amount of time the connection would be unused, make it a bad choice.

Analog versus Digital

At the risk of dredging up too much history, let's revisit the traditional telephone call. Thirty years ago, when you talked on the telephone, your voice was converted by a carbon disk from sound waves into electrical energy that was then transmitted down a wire as continuous varying voltage. That voltage stream was converted back into sound waves by another carbon disk at the receiver.

In the last twenty years, it seems as though almost all aspects of our lives are becoming computerized, digitized. Communications is no exception. Today, a typical telephone conversation is probably transmitted using digital technologies. The sound waves going into the mouthpiece get converted into numbers that get sent down the line. The receiver then converts those numbers back into sound waves that get played into the recipient's earpiece.

This transition from analog to digital in most things is now an accepted part of life. For both wired and wireless communications, digital technologies provide some excellent advantages over analog technologies. First, digital communications can be manipulated and managed by software, not hardware. That makes it possible to build much more sophisticated communications switching products. Equipment vendors can often upgrade their communications system by just loading new software. Digital communications are also less prone to interference, are more secure, and can be run at higher speeds than analog technologies.

Another key advantage to digital communications networks is that voice and data look the same. They are just streams of numbers. The network switches don't necessarily know or care whether they are managing a voice call, a WAP connection, or a Web surfing session. One set of technologies can be used for many different purposes.

As we explain in upcoming sections of this chapter, there are various competing wireless communications technologies throughout the world. The predominant systems in Europe are digital, and have been since the early 1990s. Because the United States was slower at deploying wireless communications systems, the predominant systems in the United States are analog, but are transitioning to digital fairly rapidly.

Data communications can work on either type of network, but are most effective on digital networks for the reasons cited above. If you use a wired or wireless analog network for data communications, you typically have to use a modem to create a circuit switched connection. As we explained in the previous section, the long circuit switched setup and tear down times make analog networks unsuitable for WAP-friendly connections.

Transports and Protocols

It's not easy explaining the difference between transports and protocols because the terms are often used incorrectly. To simplify matters, we'll resort to an analogy. (Any engineers reading this book will probably find this analogy laughable, but it conveys the essence of the difference.)

Suppose you have a flashlight. It's a piece of hardware that shines a light in one direction. You turn it on and off by flicking a switch. Now, suppose you want to send a message using your flashlight to someone down the block. You both agree to stand out in the middle of the street at midnight so you can send the message by turning the flashlight on and off in some sequence. Unless you also agree on some sort of message structure, however, you won't be able to send a message that the recipient understands.

In simplistic terms, the flashlight is a *transport*, a means by which you send information from one place to another. It's hardware with limited intrinsic value. The code you choose to communicate the message contents is the *protocol*. It's an agreed-upon set of standards that you can use to create higher-level value. For instance, the well-known codes for SOS, a universal distress signal, are three short dashes, followed by three long dashes, followed by three short dashes.

A transport can usually be adapted to more than one protocol. For instance, you could use Morse code to send your flashlight message. Alternately, you could use some high-security naval inter-ship code instead. As long as the sender and receiver agree on the coding, it should work.

In the world of wireless communications, each protocol typically has many different *layers*, which is why they're difficult to describe. After all, a library is composed of books, which are in turn divided into chapters, sections, paragraphs, words, and ultimately letters on paper. You could say that the paper is the transport, but which of those other things effects the communication? Protocols are similarly complex.

In wireless communications, one person's protocol may be another person's transport. More confusing, one protocol may run on top of another. Also, there are certain systems such as CDPD where the name is usually used interchangeably to identify both the transport and the protocol, and not necessarily accurately.

The ISO Network Model

Most protocols are described using the seven-layer International Standards Organization (ISO) Open Systems Interconnection (OSI) data communications model shown in Figure 1.1. This model is used to conceptually explain how various networks work. It contains seven layers that identify the most common services that you normally find in all networks, whether they are wireless or wired. These layers all interact with each other to make the network work.

The seven layers, from top to bottom, are:

Physical Layer. The physical layer encapsulates the electrical and mechanical characteristics of a medium that transports communications signals. For wireless communications systems, this is some part of the electromagnetic spectrum and the characteristics of the electrical signals that are used on that spectrum.

Data Link Layer. The data link layer defines the format of the signals that are responsible for transporting data across the medium. These are low-level electrical signals generated by a system's hardware.

Network Layer. The network layer provides services for identifying and connecting two nodes on a network.

Transport Layer. The transport layer is designed to provide reliability in a network connection such that when one device sends a message to another, the receiving device actually gets the message.

Session Layer. The session layer's job is to establish (setup), manage, and disconnect (teardown) sessions.

Presentation Layer. The presentation layer's job is to negotiate the format of the data that is sent back and forth between two devices.

Application Layer. The application layer executes a specific program such as a file transfer or e-mail exchange.

The higher layers call on the lower layers to do their job. For instance, the presentation layer on one device calls on the session layer to create a connection to another device.

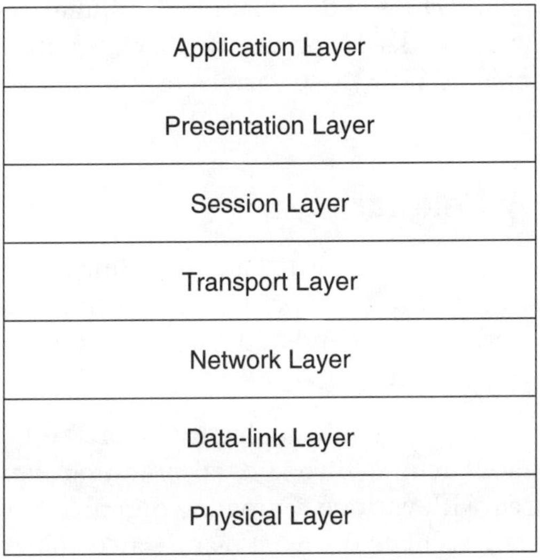

Application Layer
Presentation Layer
Session Layer
Transport Layer
Network Layer
Data-link Layer
Physical Layer

Figure 1.1 The ISO Network Model.

As we've already indicated, the terminology used to describe communications systems can be complex. For instance, an ISO-compatible network description is usually called a *protocol stack*. Within the stack, each of the seven layers may be defined by a separate protocol definition. The bottom few layers are usually identified as the transport because they are the layers that get data transported from one place to another. The upper layers tell the devices on the network what the data means and what to do with it.

Networking systems are quite complex. Our brief introduction to the ISO model is designed to give you a flavor for some of the terms you may encounter, not make you an expert. If you are interested in more details on network systems, take a look in our bibliography for some helpful references.

Wireless Technologies

Now that we've covered the key background issues, let's delve into the important wireless technologies used throughout the world. We don't include all technologies, just the predominant ones.

As we've already pointed out, it's not always easy to tell what's a transport and what's a protocol. Sometimes one industry term is used to describe both; sometimes two terms are required. That's why we call this section Wireless Technologies, instead of Transports or Protocols.

As we describe these technologies, where important, we try to differentiate between protocols and transports. If this section gets confusing, look for the handy network chart in the next section of this chapter. It summarizes the major wireless voice and data networks. Hopefully it will clarify things for you. It should help you understand how these various technologies fit together around the globe.

AMPS and European Analog Cellular

AMPS, or Advanced Mobile Phone Service, is the analog cellular transport used throughout North America and in other parts of the world, notably Central and South America, and New Zealand and Australia. It has the best coverage of all North American systems.

AMPS operates at 800 MHz. It is a voice-only analog transport. You can also use it with a cellular modem for circuit-switched data communications. AMPS is slowly being replaced with various competing digital networks. For the foreseeable future however, it will be the most readily available cellular network in North America.

At the same time that AMPS systems were being built in the United States, a variety of incompatible analog systems were promoted in Europe and the rest of the world. Although they all operated in the 900 MHz frequency range, the European systems did not work with each other.

These 900 MHz European analog systems, which we don't identify individually, are rapidly being phased out in favor of all-digital systems.

TDMA

TDMA, or Time Division Multiple Access, is a digital transport that divides the frequency range allotted to it into a series of channels. Each channel is then divided into time slots. Each conversation within that channel gets a time slot, hence the term "division" in the name.

TDMA as a transport can be compared to a country's highway system. A TDMA channel is like a traffic intersection where several lanes merge into one. In countries where the drivers are courteous and well-behaved, they take turns. Lane one goes, then lane two, then lane three, and so on. Individual connections, both voice and data, are identified by the location of their time slot.

TDMA has been in use for quite some time in Europe as the basis for the GSM (Global System for Mobile Communications) which we describe shortly. More recently, it is being adopted in North America in some PCS systems.

It's possible to overlay TDMA on top of an AMPS transport, converting an analog network to a hybrid analog/digital network. Some AMPS carriers in North America have been doing this to add security, capacity, and data capabilities to their older voice systems. This type of network has several names, including Digital AMPS (D-AMPS) and North American TDMA (NA-TDMA).

A well-known wireless company called Nextel uses TDMA technology in the SMR (Specialized Mobile Radio) spectrum block just adjacent to the 800 MHz AMPS spectrum in the United States to implement a hybrid analog/digital network called iDEN (Integrated Dispatch Enhanced Network). iDEN provides voice service, plus circuit-switch data connections, and 140-character short message services (SMS). SMS is a two-way paging service that lets you send and receive relatively small data messages, even when you are making a telephone call.

CDMA

CDMA, Code Division Multiple Access, is a digital transport that has been in use by the U.S. military since the 1940s. However, as a commercial wireless transport, it is the new kid on the block compared to TDMA and AMPS.

Pioneered by U.S.-based QUALCOMM, a CDMA transmitter assigns a unique code to each wireless connection and then broadcasts its data out on the channel simultaneously with all other connections. The receiver is able to decode each conversation by knowing the unique code assigned to each connection.

CDMA is often described as a party in a room where everyone speaks a different language. If everyone speaks at approximately the same volume, you should be able to hear all the conversations. If you know the unique code (language) used by each speaker, you can hear and understand all the conversations.

CDMA advocates claim that CDMA has some definite advantages over TDMA. First and foremost, CDMA supports more simultaneous users: approximately 10–20 times AMPS, and three times TDMA. It uses less power, giving you much better phone battery life. It is also more secure, because it hops from one frequency to another during a conversation, making it less prone to eavesdropping and phone fraud. Other benefits include fewer dropped calls and better voice quality.

CDMA is being widely deployed in North America in new PCS systems but less widely throughout the world. Like TDMA, it can also be overlaid on top of AMPS systems to create hybrid analog/digital networks.

For more information about CDMA, visit QUALCOMM's web site at www.qualcomm.com.

GSM

In the late 1980s, noting the wide disparity of analog cellular systems in Europe, various European political, trade, and academic interests started collaborating on an all-digital cellular communications network. Eventually called GSM, for Global System for Mobile Telecommunications, it has gone on to be the most widely deployed digital network in the world to date. It's used by millions of people in more than 200 countries.

Using an all-digital, TDMA-based network, every GSM phone has access to a variety of data functions at speeds limited to 9,600 bits per second (the effective throughput is typically about half that speed). These services include direct-connect Internet access (both circuit switched and packet data) without requiring a modem, mobile fax capabilities, and short message service.

GSM started out operating in the 900 MHz frequency range in all European countries. Additional networks are being deployed in the 1800 MHz frequency range. An alternate name for GSM is PCN (Personal Communication Network), the European equivalent of PCS, Personal Communication Services.

For more information about GSM, visit www.gsmdata.com.

CDPD

CDPD, or Cellular Digital Packet Data, is a TCP/IP based mobile data-only service that runs on AMPS networks. TCP/IP, which stands for Transmission Control Protocol/Internet Protocol, is the protocol underlying the Internet. Because CDPD runs on analog networks, it requires a modem to convert the TCP/IP-based data into analog signals when sending and receiving. Because of this, CDPD-friendly networks offer analog voice, circuit-switched data (made possible by the modem), and packet data services.

CDPD has a raw throughput of 19,200 bits per second. Unfortunately, the TCP/IP protocol consumes about half that, giving you an effective data throughput of about 9,600 bits per second. CDPD is designed for relatively quick set up and teardown, making it similar to packet data connections. However, it's not as efficient as digital-only networks for short, bursty data communications.

CDPD is a uniquely North American protocol that isn't widely used elsewhere in the world. In fact, it has not been widely deployed in the United States. CDPD will most likely be replaced by various all-digital networks in the coming years.

Voice/Data Networks

The previous section of this chapter describes the major global wireless data technology contenders. Table 1.1 lists each of the major cellular voice communications networks that also support data. The table lists alternate names, the type of technology (analog, digital, or hybrid), the frequency range used by the network, and the part of the world where it's predominant.

Note that the locations are either United States or Europe, often followed by the word *global*. Most of these networks are most popular in either the United States or Europe. However, they also enjoy some degree off success in other parts of the world. This is indicated by the word "global" in the location column.

A network is a unique combination of a spectrum block, a transport, and a protocol. As we've said, different networks often have multiple common names and transport and protocol names are often used interchangeably. This can make things a bit confusing.

All of these networks support circuit switched data connections. You can use circuit switched connections to access WAP data, but it's very inefficient. All of these networks, except for pure AMPS, support packet data-like connections or Short Message Service (SMS), both of which can be used for WAP.

Table 1.1 Voice/Data Networks

NETWORK NAMES	CLASSIFICATION	FREQUENCY	LOCATION
AMPS	analog	800 MHz	US/global
AMPS/CDPD	analog	800 MHz	US
CDMA	analog/digital	800 MHz	US/global
iDEN	analog/digital	800 MHz SMR	US
TDMA/D-AMPS/NA-TDMA	analog/digital	800 MHz	US/global
Various	analog	900 MHz	Europe/global
GSM/GSM 900	digital	900 MHz	Europe/global
GSM/GSM 1800/PCN	digital	1800 MHz	Europe/global
CDMA/PCS/PCS 1900	digital	1900 MHz	US
TDMA/PCS/PCS 1900	digital	1900 MHz	US
GSM-NA/GSM 1900/PCS 1900	digital	1900 MHz	US

Future Wireless Communications

In spite of the fact that the cellular communications landscape is currently a mess, particularly in North America, proponents of the various networks are hard at work on two more generations of their respective technologies, insuring that things will get even more chaotic. What's being promised is more speed.

By 2002, there should be wireless cellular networks that can provide data connections in the 50 Kbps (thousand bits per second) range. By 2005, speeds should reach up to 2 Mbps (million bits per second), letting us do such things as quickly send photographs from digital cameras to our friends and family, and receive real-time video using portable wireless devices. Whether anyone will want to do these kinds of things remains to be seen.

Like the current state of affairs, there are several high-speed wireless data technologies with names such as GPRS, CDMA2000, and EDGE that are being touted as the next wave of wireless data. Several of these systems are currently being tested in limited trials in various parts of the world. This means that the current confusing wireless communications landscape will get even more complicated as current technologies are replaced by their younger siblings.

In general, data speeds will get faster. Data connections with cell phones (or data-only devices like two-way pagers) should also get easier and less expensive. That's great news for WAP.

CHAPTER 2

A Brief History of WAP

When you think of history, your mind may wander to the Renaissance, to the Age of Dinosaurs, or to the Big Bang. With WAP you need only look back to a press release dated June 26, 1997. On that day, three industry heavyweights—Ericsson, Motorola, and Nokia—and a relative unknown—Unwired Planet, now Phone.com—announced the creation of a new technology for delivering Internet content to all types of mobile and wireless devices. Shortly thereafter, in December 1997, the four companies announced the formation of the WAP Forum Ltd. (www.wapforum.org).

In stark contrast to other technologies and markets, these companies created the WAP Forum to share information and to create an open standard. Each of the companies independently recognized the imminent convergence of voice and data communications. Because of this openness, WAP has avoided the tragic end that other technologies often encounter as companies and alliances battle to become a standard. Additionally, this openness has fostered a rapid adoption rate by the majority of handheld, paging, and cellular phone companies. In less than three years WAP evolved from an idea in four companies' minds to a worldwide industry standard currently being implemented.

Origins

While the four companies that founded the WAP Forum all had a hand in the currently available WAP technology set, its basis was a gift from Phone.com.

The company incorporated in 1994 as Libris, Inc. and has changed its name twice: first to Unwired Planet and then to Phone.com. By November 1995 the company hosted the first public demonstrations of its UP.Browser, a micro-web browser for cellular phones.

While HTML and related technologies such as JavaScript, Java, and Flash work well for desktop computers and laptops with large displays, it's a poor markup language for devices with small screens and limited resolution. Color graphics, animation, and sound challenge developers under the best of conditions. Additionally, these types of devices lack the processing power and memory to handle multimedia.

To combat this, Phone.com developed a set of technologies related to HTML, but tailored to the small screens and limited resources of handheld, wireless devices. Most notable is Handheld Device Markup Language (HDML). HDML on paper looks similar to HTML, but has a feature set and programming paradigm tailored to wireless devices with small screens. HDML and other pieces of this architecture eventually become Wireless Markup Language (WML) and the architecture of WAP. See Chapter 4, An Overview of WAP, for more details on the WAP architecture.

Between November 1995 and June 1997, Unwired Planet negotiated major contracts with many prominent cellular phone makers to use their HDML-based UP.Browser, and with cellular phone infrastructure companies to install UP.Link Servers to handle requests from the UP.Browser. Mitsubishi demonstrated the UP.Browser running on their Mobile Access Phone in January 1996. AT&T Wireless, Bell Atlantic Mobile, Samsung, QUALCOMM, and GTE quickly followed with announcements that they too would utilize Unwired Planet's technology.

In June 1997, Unwired Planet, along with Ericsson, Nokia, and Motorola, announced the formation of the WAP Forum. Instead of fighting imminent competition from other companies offering their own standards, these companies sought to make their technologies the standard for mobile Internet access. Unwired Planet offered HDML, the markup language, and the Handheld Device Transport Protocol (HDTP); Nokia brought their Smart Messaging protocol; Ericsson offered their Intelligent Terminal Transfer Protocol (ITTP). This alphabet soup simmered for a few months until April 1998 when the Forum delivered the WAP 1.0 specification.

The specification is a set of documents describing the protocol. There are several, they're long, and they're technical. They cover everything from the overall architecture and security information to the binary format of a WAP application and a description of WMLScript (similar to JavaScript). The documents contain enough information for any developer to learn the minutia needed for creating WAP-based products.

Since its initial release, WAP has evolved twice. The current version is 1.2 and without a doubt version 1.3 will follow to accommodate the changing characteristics of WAP devices and wireless networks entering the market. Table 2.1 summarizes the iterations of the WAP specification. Both releases 1.1 and 1.2 of the specification have the same functionality as 1.0, but add features to align with what the rest of the industry is doing.

> **Japan's i-Mode**
>
> While America dreams of tiny, smart phones with large displays, the ability to send and receive photographs from a handheld device, and instant access to any type of information, Japan lives the reality. Japan boasts one of the largest cellular and smart phone subscriber bases in the world. Of its 126 million residents, 40 million, or 1 in 3 people, carry a cellular telephone. Nippon Telephone and Telegraph (NTT), Japan's version of AT&T, owns a 67 percent stake in one of Japan's hottest technology companies, DoCoMo. DoCoMo created a potential WAP competitor called i-Mode. It gives users access to specially formatted web sites and portals that contain news, weather, sports, horoscopes, and cartoons, and lets users trade instant messages, e-mail, and photographs taken with the phones. To date, with more than 4,000 specially formatted sites and millions of subscribers, i-Mode has brought to light the apparent need for smart phone-based services. (Kunii & Baker 2000, p.88.)
>
> i-Mode is a 9,600 bps packet switched service. Packetswitched networks send and receive information quickly, without the need to establish a connection the way a traditional modem does in a circuit switched environment. While not high in bandwidth, i-Mode can seem almost real time to the users. DoCoMo's two largest competitors are DDI and IDO of Japan, both of whom are backing WAP. NTT DoCoMo has joined the WAP Forum. It remains to be seen how they will support WAP in the future, but the fact that they are members further legitimizes the standard.

While a significant amount of technology from Phone.com led the charge for the WAP 1.0 specification, it was only with the help of the other founders that WAP has been so well accepted.

Table 2.1 WAP Specification Summary

VERSION	RELEASE DATE
1.0	April 1998
1.1	June 1999
1.2	November 1999

Ericsson. Ericsson, headquartered in Sweden, is currently a world leader in mobile communication systems. Ericsson equipment carries 40 percent of the world's cellular phone calls whether it's the phone itself or the underlying infrastructure carrying the call. Their technology embraces all of the major digital cellular technologies: GSM, CDMA, and TDMA. Revenues for Ericsson in FY 1999 were almost $25 billion USD.

Motorola. Motorola, an Illinois-based company, is another major manufacturer of cellular phones. Additionally, they create a wide range of integrated communication solutions and embedded systems. Their proficiencies include semiconductors and all types of two-way, paging, and satellite networks, and Internet access products. In fact, you would be hard-pressed to find a consumer electronic product that does not have a Motorola microchip inside. Motorola's revenue for FY 1999 exceeded $31 billion USD.

Nokia. An Espoo, Finland based company, Nokia sells more cellular phones than any other company in the world. They offer products that work all over the world and that compete with Motorola and Ericsson. Revenues for Nokia were almost $20 billion for FY 1999.

You can find more information on each of the founding companies and their WAP initiatives at the following web sites:

- Ericsson—www.ericsson.com/wap
- Motorola—www.motorola.com/MIMS/MSPG/spin/mix/mix.html
- Nokia—www.nokia.com/wap
- Phone.com—www.phone.com

The WAP Forum

The WAP Forum's purpose is to foster the growth of the WAP protocol through an open standard available to all. It encourages participation from all facets of the global telecommunications industry. Members are asked to contribute to the standard so that the solutions they create through consensus work for everyone.

Specifically, the Forum's goals include:

- Fostering the development of advanced wireless services for consumers and the enterprise. The services can provide, for example, news, weather, e-mail, and electronic commerce. Also in development are a number of enterprise and vertical markets solutions for business. These wireless applications and services provide a new avenue for mobile workers to stay in touch with the office and corporate data and allow business-to-business transactions to happen.

- Creating a global protocol that is available to and compatible with all wireless networks. The intent of this is for WAP to remain network independent. The fewer ties it has to a particular technology, the more likely WAP is to work for the widest range of devices and networks. The WAP Forum wants to avoid situations where WAP works for some phones on some networks in certain geographic areas. The WAP Forum's openness has created a global following ensuring its success as the standard for wireless data transfers to the Internet.

- Enabling the creation of applications that scale across a variety of networks and devices. The WAP standard lets hardware with different screen sizes, user interface controls, and processing abilities use the same application on all devices. This is not unlike what Sun Microsystems is trying to do with Java. You create a WAP application once. It should run on any WAP device from any manufacturer without alteration.

- Utilizing and extending existing Internet standards. WAP leverages current Internet technologies like HTTP, HTML, and JavaScript to work its magic. This helps avoid competing standards and lets current Internet developers quickly design new applications and migrate existing ones for use on wireless handheld devices.

These goals are lofty and would be for naught if the WAP Forum did not gain the support of other technology companies. In the three years since its inception, the Forum has added over 350 members to its ranks. The members include cellular phone makers, cellular network providers, and software designers from all over the world.

Forum Members

The acceptance of WAP by companies has been rapid to say the least. In addition to the four founding members, more than 350 companies are listed as members of the Forum (as of April, 2000). The members are a who's who of the communications world. Cellular infrastructure, network, software, online service, and consulting firms have all joined. Each plays a role in making the hardware, carrying the calls, creating the software, and offering the services that will make WAP a ubiquitous standard for everyday people using data services while on the road.

In the following sections, we describe various types of companies and list some representative companies by name. Please note that these lists are far from complete, and that the wireless communications community changes very rapidly. For the most complete, up-to-date list of WAP Forum members, refer to the WAP web site at www.wapforum.org.

Hardware Providers

The hardware providers are the companies that come to mind most readily when you think of cellular technology. These companies create the telephones and handheld computers, determine their features, and design the user interfaces you've come to rely on everyday. Members of this group include Ericsson, Motorola, Nokia, PageNet, Palm Computing, QUALCOMM, Symbian, and Sony.

Cellular Service Providers

Cellular service providers comprise the second group of WAP supporters. They frequently appear on television, radio, and in print selling a certain number of minutes per month for their phone plans. They decide on additional features, including the WAP services available to subscribers. This group contains AT&T Wireless, Bell South Wireless, NTT DoCoMo, and IDO.

WAP Infrastructure Creators

WAP is not simply a hardware-based solution. WAP requires two key pieces of software: WAP browsers for viewing content on smart phones and handheld devices, and WAP gateways. WAP gateways sit between the mobile device and Internet web server converting WAP requests to HTML requests and translating the resulting response into a compressed binary form for delivery to the mobile device.

Several companies have created WAP browsers for smart phones and handheld devices. They have different capabilities in terms of rendering text and graphics and executing WMLScript, much like Netscape Communicator and Internet Explorer differ. WAP gateway manufacturers will inevitably create software that varies in capability and performance as well. Providers include Ericsson, Nokia, Phone.com, Motorola, NTT DoCoMo, and WAPIT Ltd.

Software Developers

The next piece of the WAP puzzle includes the makers of web servers, server-based applications, and databases. The current set of server-based software for the web almost exclusively outputs HTML pages. With only a few exceptions, the entirety of web content viewing is done through traditional web browsers on desktop and laptop computers. However, many of the medium and larger companies have embraced the potential of the mobile, handheld, and smart phone markets and are scrambling to convert their products to WML and WMLScript. Software developers include Microsoft, Oracle, Allaire, and Sybase, among many others.

The second area of software development includes consultants and systems integrators. As with any new technology, business always finds ways to leverage it to reduce cost, increase productivity, and work in ways not possible before. The emergence of WAP will create companies with strategic expertise with WAP, and system integrators who can apply WAP tools to solve business problems.

Content/Service Providers

No service would be complete without information providers. They need to embrace the format and create information services tailored to devices with limited screen real estate, memory, processing power, user input capabilities, and bandwidth. While challenging, no one can dispute the hundreds of millions of cellular phones being sold worldwide and the popularity of the Internet. Just as e-commerce has redefined retail, WAP will help create a mobile e-commerce market. Though this is the last group described, it will undoubtedly encompass the most companies. Current providers are America Online, Amazon.com, Charles Schwab, MapQuest, ShopNow.com, MasterCard, and Visa.

Available WAP Services

Although the number of announced WAP products and services is exploding, WAP deployment is in its earliest stage around the world. Numerous cellular service and content providers are teaming up to let customers try out WAP services on smart phones. While WAP is gaining steam, you'll also find a healthy number of service providers in the United States still using Phone.com's HDML technology. Over the next 2-3 years you'll find HDML services remade into WAP along with hundreds, if not thousands of new WAP services.

While this area of the market is extremely volatile as new services come on line daily, there are a few worth noting. You'll find big name players already in the market sporting both HDML and WAP options. This type of support demonstrates the faith companies have in WAP as a worldwide standard. While the list is current at the time of writing, expect hundreds more by the time you read this.

Sprint PCS Wireless Web. Sprint PCS offers a wide range of (currently) HDML services for Sprint PCS users who own WAP-enabled cellular phones. Content providers include Bloomberg.com, CNN, Ameritrade, InfoSpace.com, Yahoo!, and MapQuest.com.

Verizon. The merged GTE and Bell Atlantic companies in combination with Vodaphone-Airtouch's cellular division are offering a set of HDML services, branded Mobile Internet, to cellular subscribers. These services include e-mail, stock viewing, news, weather, and access to Amazon.com to purchase books, music and video games.

Bertelsmann AG/America Online. These two companies, along with help from Nokia, Ericsson, and RTS Wireless, are working on delivering WAP-formatted AOL content to England, France, and Germany.

Yahoo!. One of the ubiquitous names on the Internet isn't sitting by the wayside. It has WAP portals up and running in a number of European countries offering the usual news, weather, and sports. Yahoo! also offers access to My Yahoo! personal pages.

The WAP juggernaut is nothing short of remarkable. Within two years, practically everyone in any industry related to wireless communications has become aware of WAP. There are predictions that within the next few years, mobile computing and cellular telephony will incorporate WAP into the core of their business models, affecting hundreds of million of end users around the world.

CHAPTER 3

Why WAP?

As we enter the 21st century, mobile technologies such as cellular telephones and handheld devices for tracking personal information are becoming a staple rather than a luxury. A few years ago you wouldn't dare leave your house without a wallet or purse, keys, identification, and some sort of money. Today that list, for millions of people, includes a cellular telephone, and soon, a smart phone with Internet access. Because people demand data access while mobile, they need a technology that can deliver the information in a format suitable for devices with small screens and limited memory, processing power, and bandwidth. Enter WAP, the Wireless Application Protocol.

The Great Convergence

WAP was not created to force everyone to purchase new cellular telephones and pay for additional data services. Rather, its emergence comes as a logical conclusion to the convergence of three existing technologies coupled with strong social and economic forces.

The Internet

What can you say about the Internet? Some compare its impact on the world to television. Others say that its influence goes well beyond and we don't yet understand all of its ramifications. Nevertheless, the Internet is redefining how we communicate, learn, buy and sell goods, and share. The information in most cases is free, while the volume in almost all cases is overwhelming.

When the Web first appeared, web sites were text based, the number of pages was limited, and the information scattered. Over the past five years, that's all changed. Large, coherent sites specializing in hundreds of topics are the norm. They're enriched with multimedia and let users customize their site usage. For example, My Yahoo! lets users tailor a web page to include news, weather, sports, stocks, movies, and other topics that are of interest to that individual user. The goal is to create a personal experience that compels the user to visit time and again. The result is that people rely on the Web for information. Unfortunately, the vast majority of web surfers are relegated to experiencing it from a desktop computer hardwired to the wall.

Thankfully, all that is changing. Currently, a number of incompatible technologies exist which provide wireless data access to the Internet. In the United States, this comes in the form of news and information services running on Phone.com's HDML (similar to HTML) technology, two-way paging networks, cellular modems, and the like. In the summer of 1999, Palm, Inc. attempted to remedy this chaotic situation with the Palm VII, an out-of-the-box wireless device. It serves as a good example of what's to come with WAP.

The Palm VII solves a number of problems with wireless Internet access. It is the first product that contains a large screen (relative to cellular telephones and pagers), has a reasonable amount of computing power, and runs on a network (BellSouth Wireless) with excellent coverage. Palm believes that wireless data communications should be painless for the end user. By creating the hardware and providing the service through Palm.net, Palm created one-stop wireless shopping for their customers. No longer did a user need to call an operating system manufacturer, cellular modem manufacturer, and an Internet Service Provider to find exactly why they could not connect and retrieve data. Palm VII has a user up and running within minutes of opening the box.

The Palm VII does not contain a general-purpose web browser. Palm is realistic about the abilities of small screen devices that lack mice and keyboards. They use a technology called Web Clipping to take subsets of data from the Internet for presentation on a Palm VII. Web Clipping uses a tailored version of HTML to create content fit for small screens and limited bandwidth. Developers create Web Clipping applications that do a specific task, such as retrieve news or find the nearest Starbucks coffee shop. The Palm VII has generated a significant following since its launch. For a relatively small amount of money, people have quick, easy access to Internet content from hundreds of cities.

Palm is a member of the WAP Forum. Although it's not currently clear how they plan on supporting WAP, the Palm VII gives us a taste of what to expect from WAP.

Wireless Technology

Wireless connectivity continues to make monumental improvements in coverage and performance every few months. While Europe and Japan enjoy pervasive digital networks, America plays catch-up. Analog represents the majority of cellular phone calls in the United States, and is limited by interference, eavesdropping, and low-speed data rates (19.2 Kbps nominal, 9.6 Kbps actual).

Today, cellular carriers and major communication corporations like Sprint, MCI WorldCom, Ericsson, NTT, and AT&T spend billions of dollars around the world to create and upgrade networks to support incompatible digital voice and data communications and expand their coverage. In the United States it is common to have digital coverage in one city, but lose it in the next due to an incompatible network. Worse if you leave the country, your cellular phone has a negligible chance of working at all.

The growth of the cellular telephone market is nothing short of astounding and the facts are staggering. There are more than 300 million cellular phone users in the world. Cellular telephone manufacturers expect to ship an additional 300 million devices in 2000 and an additional one billion telephones by the year 2003. Estimates on the number of telephones that will be WAP-enabled range from 30–80 percent. Table 3.1 summarizes cellular telephone sales for the next four years according to market research firm DataComm Research.

Computing Power

One of the most commonly mentioned quotes in the computer world is Moore's Law. It states that computer processing power doubles every 18 months. Since the 1960s, when Gordon Moore, former chairman of Intel, predicted this performance gain, the law has held true. The doubling of computer power has forced manufacturers to find processes for creating smaller chips on which they can pack more functionality. As the components of the chip shrink, so do their power requirements. The results, year in and year out, are successive generations of microchips with increased performance and decreased power consumption.

Chip manufacturers have also spent significant effort integrating this general purpose computing power with audio, video, and voice recognition tech-

Table 3.1 Cellular Phones Sales (Millions of Units)

2000 DATACOMM RESEARCH	1999	2000	2001	2002	2003
Smart Phone Sales	.2	6	56	175	330
Conventional Cellular Phones	250	304	284	185	40

nologies. The results are chips that can handle all types of computing tasks and can still fit inside the small space of a cellular phone. The integration of microprocessors and operating systems into cellular phones makes them "smart" and will be the means by which you search for information, send e-mails, read news, and interact with the Internet in general. Long term, companies will exploit powerful microchips, new display technologies, and audio and video processing to give us smart phones that let us videoconference in full color, watch television, listen to audio, send and receive photographs, and communicate with other types of devices.

Social and Economic Forces

Simply creating technology for its own sake is not enough to create demand for a product. Social forces, shown conceptually in Figure 3.1, have to contribute as well.

- People in general are becoming more mobile. Low cost airfares for vacationers and an emphasis on global business have more people traveling to a wider range of locales. Whether or not you had access to information was simply a matter of whether or not you were in the office. Now companies spend billions of dollars on wired and wireless technologies to keep their workers in touch and productive.
- Because of the Internet, people are becoming information junkies. Knowing the next show time for a movie, the current price of a stock, or being able to look up corporate information for a client typically ties someone to a computer at a desk, to a television, or to a landline phone. The dramatic increase of mobile workers, at-home workers, vacationers, and business travelers in a global economy requires that data be available while mobile.

For these reasons, people need guaranteed information access away from the home. Regardless of the device type, the information should be ubiquitous and easily accessible from anywhere in the world.

It should also be cost effective. Fortunately, other forces are at work which make that a reality as well:

- The cost of buying and using cell phones is decreasing. In numerous areas of the world it is less expensive to use wireless voice communications than traditional landlines. Some less industrialized countries that are installing their first communications infrastructure are opting to skip landlines and make a wireless network the standard for cost reasons.
- Internet access in all countries and cellular communications in Europe and Asia have used a flat rate pricing model for a few years now. In the

Why WAP? 25

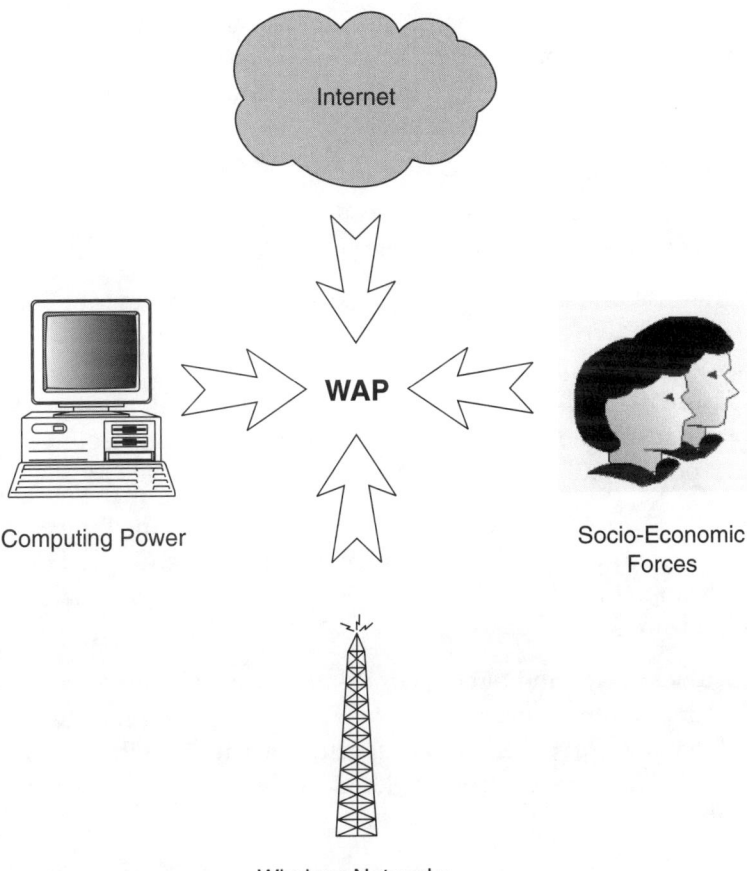

Figure 3.1 The need for WAP.

United States, cellular phone pricing schemes are heading in the same direction. This will be considered the norm in the future. Flat-rate pricing makes cellular telephones attractive to many people who once thought of them as a luxury.

WAP Device Characteristics

Traditional software publishers frequently list exacting specifications required to run their software. They might require, for example, a Pentium III processor, 64 MB of RAM, and 200 MB of free hard disk space.

The WAP Forum defined a general list of device and network characteristics that is intentionally subject to interpretation. The standards take into account the variety of screen sizes, processing power, bandwidth issues, and input methods or dissimilar wireless devices.

First, they stated that a WAP device has limited CPU power, RAM, and ROM. Exact numbers are not given in the specification. Limited simply means enough to get the job done, but nothing more.

Second, they assume limited battery life. Batteries for cellular phones range in duration from two to as many as 10 hours. Battery life continues to improve, but, unlike processing power, at an incremental rate. For the foreseeable future this will be true and the WAP specification must work under these conditions. If the specification required a device that required a large amount of power, battery life would suffer.

Third, WAP networks have limited bandwidth. Most of the time we're talking in the neighborhood of 10 Kbps, which is one-fifth the speed of most modern consumer modems. 3rd Generation wireless networks will correct this, but they are several years away.

Fourth, wireless voice and data networks are unreliable. Frequently, users experience interference, lost connections, and interrupted service. The WAP Forum designed WAP to be a robust specification that can handle these types of conditions. This prevents application designers from having to worry about such issues.

Fifth, WAP devices have small screens and limited input capability. It just isn't practical to expect users to use a telephone keypad for data entry in the same way they use a QWERTY keyboard. Cellular telephone manufacturers each have their own design methods for input and screen size.

The Need For WAP

In this chapter we've explored a variety of converging technologies and forces. The time is right to deliver data to mobile devices. A question remains: What technologies are most appropriate for delivering that data?

The most obvious answer is to simply use existing Internet technologies. Before WAP, people were able to get Internet access from wireless devices. Unfortunately, the standard Internet technologies have several flaws making them unsuitable for wireless access:

- Web pages as we currently know them are fine for desktop and laptop computers with large, color screens, gobs of processing power, and bandwidth 5 to 1000 times higher than a typical wireless connection. Smart phones and handheld devices have small screens, modest processing power, low speed wireless connections, and limited input abilities. Typical web pages just don't work with those types of devices.
- HTML, the language used to create web pages, is a text-based language which wastes a significant amount of bandwidth. It takes more time and bandwidth to send data with HTML than with binary-encoded, compressed information. The same is true of TCP/IP, the protocol used to deliver web pages to desktop computers. It wastes bandwidth.
- Current Internet technology uses TCP/IP, which does not have robust technology for handling the interference and lost signals commonly found in the wireless world.
- TCP/IP is not inherently secure. Transmissions can be intercepted and read by others.
- TCP/IP is not supported by all wireless networks.

The WAP specification was designed to accommodate all of the features, advantages, and disadvantages of Internet access through a wireless connection. WAP is:

Tailored for small devices. As we explore in a later chapter, you can easily create WAP applications that accommodate a wide range of devices.

Binary encoded. The connection between a smart phone and a WAP gateway is encoded and compressed to maximize throughput on low bandwidth networks.

Robust. WAP's design accounts for the challenges of wireless connections.

Secure. Data to and from the WAP device is encrypted, preventing eavesdropping.

Independent. WAP's design makes it device and network independent. WAP operates across a wide range of devices and wireless networks.

CHAPTER 4

An Overview of WAP

After reading three chapters on the history of wireless communications, WAP, and the need for such a protocol, it's time to take a look at WAP itself. We show you some simple sample applications, explore the WAP transaction model and compare it to the Internet's transaction model, look through the WAP protocol stack, and explore how applications are developed.

WAP in Action

In its most fundamental form, WAP lets you interact with Internet-based content using a WAP-enabled device. WAP was designed to deal with the intricacies of wireless data exchange on devices with small screens, low power, and limited input. One of the first things we should do before introducing and exploring the WAP programming model is to take a peek at what a WAP application looks like running on a device.

NOTE
To get our pictures for this example, we used Ericsson's Application Designer, a software development kit (SDK) for creating, debugging, and simulating WML and WMLScript applications on Windows 95, 98, and NT computers. There are several other WAP SDKs. Two of the more popular ones come from Nokia and Phone.com. If you're interested in trying your hand at creating a WAP application, you can find information on these developers' kits at the following web sites: www.ericsson.com/developerszone/, www.nokia.com/wap/development.html, and updev.phone.com/.

Figure 4.1 A simple WAP application.

The application in Figure 4.1 is extremely simple. It shows the title "Scott's Application" with some text below it.

The telephone has "Yes" and "No" buttons and arrow keys for navigating from one screen to another. Although the screen does not have any hints as to whether or not you can interact with the application, you can.

The "Yes" button brings up the options menu shown in Figure 4.2. The WML pages designed for this demonstration define the "OK!" and "Help" options. The "prev" option is a standard WAP choice. It lets the user go back and view previous pages much like the "Back" button on most web browsers. To navigate this menu you press the arrow keys to change your selection and then the "Yes" button to choose your selection.

Figure 4.3 shows the last screen of this simple WAP program. If you choose the "OK!" option from the menu shown in Figure 4.2, you are taken to the screen shown in Figure 4.3. This screen is of particular interest because it shows the flexibility allowed in WAP implementations.

The Web supports many types of text including **bold**, *italics*, and underline, among others. WAP does too, but doesn't dictate how text styles must be implemented. In fact, WAP microbrowsers don't need to implement styles at

An Overview of WAP 31

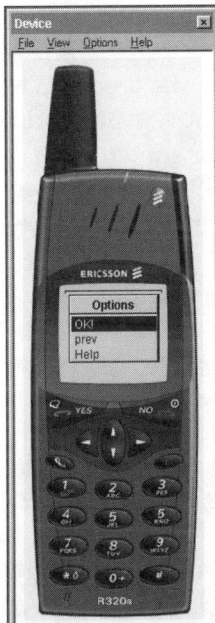

Figure 4.2 The options screen.

Figure 4.3 Various WAP styles.

all. In Figure 4.3 there are three statements on the screen in addition to the title. We tagged "This is big" to appear as Big letters, "This is emphasized" as Emphasized text, and "Italics" as Italicized text. These are three of WAP's style settings. Note that the microbrowser built-in to this Ericsson phone does not distinguish between Big and Emphasized, and Italics is shown as normal text. These are design choices that Ericsson made. The WAP specification defines these different styles, but does not dictate if and how they are implemented.

> **NOTE** Besides the microbrowser, the WAP specification makes few assumptions about a particular WAP device's user interface. The Ericsson phone shown in Figure 4.3 has four arrow keys and "Yes" and "No" buttons. Other models have buttons for up and down navigation on the side of the unit or a built-in wheel for choosing options. These differences don't require that developers create different programs for each device, but they do make the retrieval of the same information from different phones challenging.

To contrast this with other phones, Figure 4.4 shows the second and third screens of our demonstration program on a Motorola i1000 Plus WAP phone running on Phone.com's UP.Simulator which is included with the Phone.com WAP SDK. Figure 4.4 shows both screens of the program. As you can see in

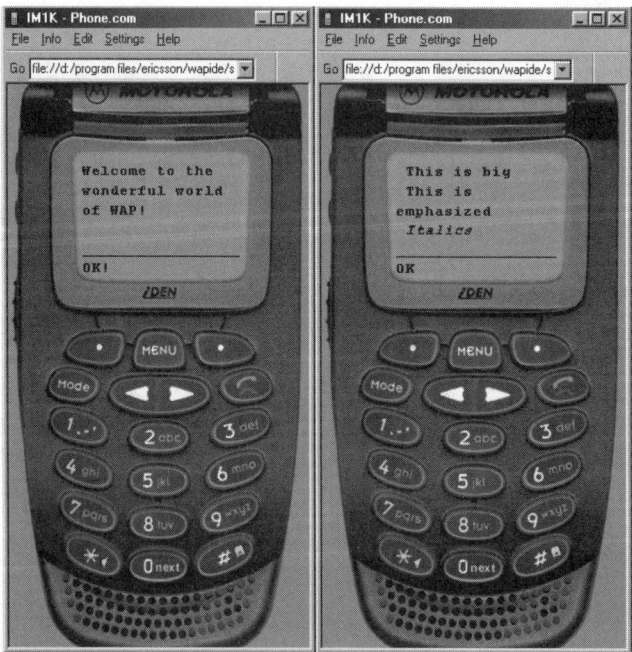

Figure 4.4 The same demonstration application running on a Motorola i1000 Plus.

the left image, the title is not visible and the "OK!" option appears at the bottom of the phone near the button you press to select that option. The right image does not show the title either, but italicized text is supported.

As you use WAP telephones with different screen sizes and features, you begin to understand some of the challenges (or benefits, depending on your perspective) of WAP. This small application runs very well on both phones. The exact implementation of the program, however, is different from phone to phone. WAP is designed to adapt to the particular features of each device.

Web Transaction Model

We've seen a simple application running on a smart phone. We said previously that WAP leverages existing Internet technology. Let's investigate how a typical Web transaction on the Internet occurs before delving into WAP transactions.

On the Web, a *transaction*, the interaction between a browser and server, starts when you type a Uniform Resource Locator (URL)—the name of a document located somewhere on the Web—into a web browser, and hit the Enter key. If the document exists, it appears in the browser's display area. If it doesn't exist, the web server returns an error page. Figure 4.5 shows the URL "http://www.wapforum.org" being displayed with Internet Explorer 5.0.

A URL has two parts. The first is the protocol. It defines how the messages transferred between the web browser and web server are transmitted and formatted. The protocol for this request is the Hypertext Transfer Protocol (HTTP).

There are several other common protocols that web browsers support. For example, HTTPS is the secure version of HTTP. HTTPS encrypts all of the data transferred over the Internet between the browser and server. FTP (File Transfer Protocol) is another common protocol. It defines how files are transferred between networked computers.

Figure 4.5 A URL in a desktop web browser.

The second part of the URL is either an Internet Protocol (IP) address or a domain name. You've probably seen numbers like 192.35.112.30 when using the Internet. Those numbers uniquely identify each computer on the Internet. Unfortunately, humans just aren't designed to remember numbers. We like names much better, so we identify URLs using names like www.nokia.com instead.

> ### What's A Protocol, Anyway?
>
> The word protocol is used throughout this book (it's even part of the WAP acronym, Wireless Application *Protocol*, so it's worth defining). In Chapter 1, we defined a protocol in somewhat technical terms. In layperson's terms, a protocol describes a format for transferring data between two devices.
>
> Let's pretend you're buying movie tickets. There is a protocol for getting those tickets. You step up to the counter and the attendant asks which movie you're there to see and what showing you're interested in. You respond with that information. The attendant then asks how many tickets you would like. You respond, two adults. You are handed the tickets, at which time you give him the necessary payment. The last step of the protocol is a "Thank you. Come again," if you're lucky.
>
> You have to follow this protocol to obtain the tickets. When the attendant asks you for the number of tickets, if you respond with Los Angeles, the protocol breaks down. Interactions between two devices using a protocol is very similar, but much more exact. Usually, most aspects of the protocol cannot be omitted or mistimed.

Web browsers aren't the only programs that can initiate valid web transactions. Any program that can create a TCP/IP connection and send a valid protocol request can get data from a web server. In fact, this is how many commercial software packages register themselves without needing a web browser after being installed on a computer.

Many times HTML files aren't the only types being requested from a web server. Files ending in "RAM" trigger RealPlayer if it's installed on your computer. A request for a "PDF" file fires up Adobe's Acrobat Reader. Many web sites today use a technology called Common Gateway Interface (CGI) to transfer data between a web server and a program running on the server. For example, if you've ever been to a web page that let you search the site's contents, you've probably experienced CGI. You enter a search word into a form and click the Search button. The web browser then uses a URL with your search term as part of the URL to access a CGI program on a web server. That CGI program queries a database for the information, dynamically compiles the results as an HTML page, and returns it to your web browser. Figure 4.6 summarizes the information flow in a typical web transaction.

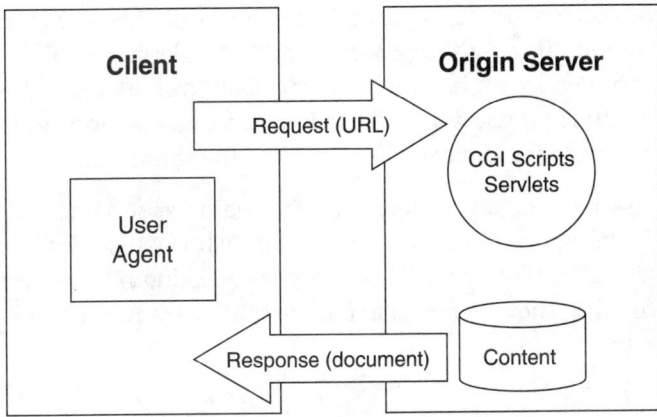

Figure 4.6 The World Wide Web transaction model.

WAP Transaction Model

We stated earlier that WAP is based to a large extent on existing World Wide Web technologies. As we explore the WAP transaction model, shown in Figure 4.7, you'll find that it is fundamentally the same as the web transaction model, but with a few key differences.

The most significant difference is the need for what's called a *gateway* between the client and the web server, which contains the information you're interested in accessing. The gateway's duties include the translation of WAP formatted messages received from the WAP device into HTTP messages that can

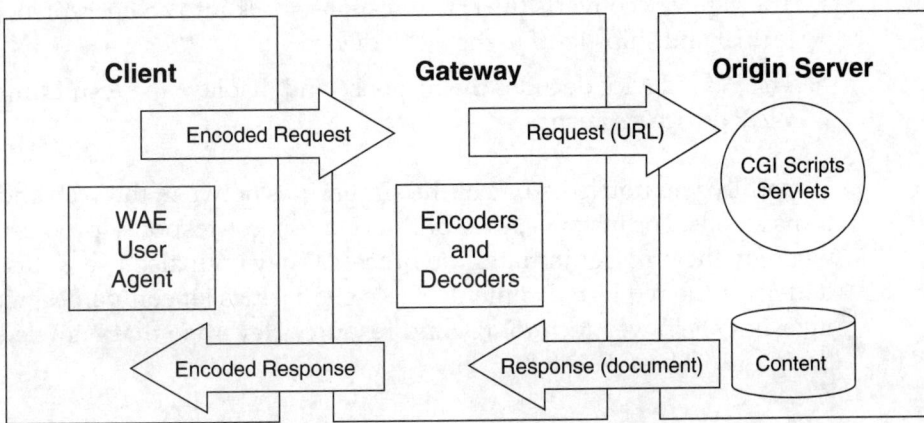

Figure 4.7 The WAP transaction model.

be sent to any web server on the Internet. When the web server responds, it will most likely send a file containing WML and WMLScript, the WAP equivalents of HTML and JavaScript. It is the gateway's job to change that text file into a WAP binary file and encrypt it. A file in this format is more suitable for wireless transmission to the device that requested the information.

The gateway is also responsible for knowing the character sets and languages of the WAP devices that use it. Whether it's an English WAP device talking to a German web server or a Japanese WAP device requesting information from a French web server, the gateway needs to ensure that the requester receives a coherent message.

WAP Step-By-Step

Let's walk through a typical WAP transaction so you understand the steps involved in retrieving information from the web server:

- A user requests a URL by entering it into a WAP device. (Alternately, an already-running WAP program requests a URL on behalf of the user.) For the sake of argument, let's say the request is for www.wmlserver.com/myweather.wml.
- The WAP device encodes the request into an encrypted, compact binary format suitable for transmission over a wireless link, and sends it to the WAP gateway.
- The gateway examines the message, converts it into a valid HTTP-based URL request, and forwards it to www.wmlserver.com.
- When wmlserver.com receives the request, it fulfills it by returning the requested document back to the gateway.
- The gateway converts the HTTP response back into an encrypted, binary format and ships it off to the WAP device.
- The WAP device decodes the response and displays the results on the WAP device's screen.

Hopefully you noticed some of the similarities between the Web and WAP transactions. For instance, they both use a request-response process whereby the browser initiates the process. They both also use web servers to deliver the requested content. These similarities let companies with investments in web technology and resources leverage that knowledge to design WAP-based systems.

WAP Architecture

Now that you've seen how WAP works under the covers, let's look at its architecture.

Figure 4.8. shows the Wireless Access Protocol as a series of layers. This layered format mimics the International Standards Organization (ISO) Open Systems Interconnection (OSI) network model, which we briefly described in Chapter 1. The OSI Model defines a layered framework for generically describing and designing protocols. The OSI Model has seven layers. WAP uses six, but the approach is similar.

Each layer in Figure 4.8 is responsible for managing some part of WAP. Additionally, each layer is only allowed to interact with the layer above and below it. This helps to define clear roles for each layer. URL requests from a WAP device start at the Application Layer and get processed until the request goes out over a Bearer network to the gateway. Responses enter the device at the Bearer level and are transformed back and finally displayed to you at the Application Layer.

While Figure 4.8 may appear overwhelming, let's look at a real world analogy, sleeping and getting ready for work, to clarify how the layered model operates.

Your body represents the information that needs to be transferred (prepared) for work in the same way a URL request is the information that must be sent to a web server. The different layers in this scenario are your bed, the bathroom, your closet, and the kitchen, in that order.

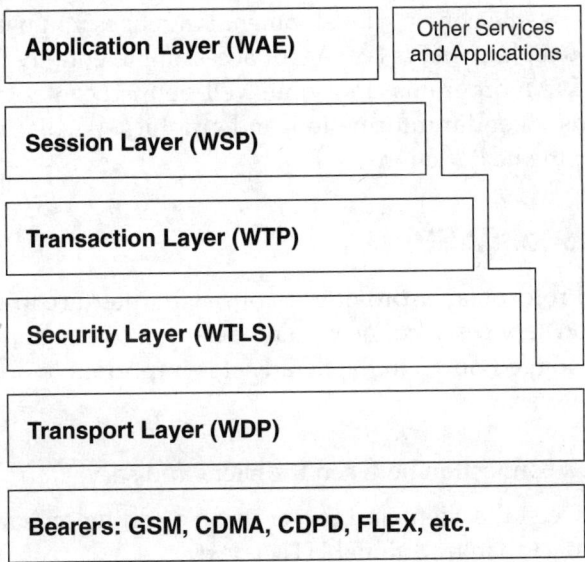

Figure 4.8 The WAP architecture.

You go to the bed layer for eight hours of sleep. The bed helps you rest. As your alarm sounds and you stumble to the bathroom for a shower, the bathroom layer is the one that helps you clean up. The closet layer helps you get dressed and the kitchen layer helps you eat. It's now time for work.

You've been transformed in each of the layers, in the right order, without skipping one, to get ready for work. If you skip the bed layer, you'd fall asleep at work. Bad idea! If you skip the bathroom layer, people at work would probably sit at the opposite end of the conference table from you.

What about doing this process in the wrong order? Dressing before showering is not a good look for work; neither is eating before sleeping because you'll be hungry at work. Bear in mind that this analogy isn't 100 percent valid because you don't reverse this process when you get home from work to get ready for bed, necessarily. Hopefully, however, you see the point.

WAP requests from an application must be transformed into a certain format before being sent wirelessly to a gateway and finally off to a web server to have the request fulfilled. The response on the return trip is unencrypted and decoded before being displayed on the screen. Each request and response must proceed through the set of layers in Figure 4.8 in the correct order each and every time.

With all that in mind, let's look at the WAP architecture layers.

WAP Application Environment (WAE)

The Wireless Application Environment layer is the one you're mostly likely concerned with if you are considering deploying WAP applications. It encompasses the devices, the content-development languages you use (WML and WMLScript), the telephony APIs (WTA) for accessing telephony functionality from within WAE programs, and some well-defined content formats for phone book records, calendar information, and graphics. We discuss this layer in more detail in the next section.

Wireless Session Protocol (WSP)

The Wireless Session Protocol layer provides a consistent interface to WAE for two types of session services: a connection mode and a connectionless service. Without getting bogged down in the details, it is important to note the services this layer provides:

- Create and release a connection between the client and server.
- Exchange data between the client and server using a coding scheme that is much more compact than traditional HTML text.
- Suspend and release sessions between the client and server.

Wireless Transaction Protocol (WTP)

Now we're getting a bit more technical. WTP provides transaction services to WAP. It handles acknowledgments so that you can tell if a transaction succeeded. It also provides retransmission of transactions in case they are not successfully received and removes duplicate transactions.

WTP manages different classes of transactions for WAP devices: unreliable one-way requests, reliable one-way requests, and reliable two-way requests. The definition of unreliable and reliable is what you might think. An unreliable request from a WAP device means that no precautions are taken to guarantee that the request for information makes it to the server.

You might think that this is a ludicrous transaction type. Why would anyone request something, but not care if it was actually fulfilled? One-way paging networks work in this fashion. If you page someone and the pager is off or out of range, that person does not receive the message. While WTP does provide this service, most transactions use one of the reliable transaction types.

Wireless Transport Layer Security (WTLS)

Wireless Transport Layer Security provides services to protect your data, including data integrity, privacy, authentication, and denial-of-service protection. Data integrity guarantees that the data that is sent is the same data that is received. WAP privacy services guarantee that all transactions between the WAP device and gateway are encrypted. Authentication guarantees the authenticity of the client and application server. Finally, denial-of-service protection detects and rejects data that come in the form of unverified requests.

Wireless Datagram Protocol (WDP)

The Wireless Datagram Protocol provides a consistent interface to the higher layers of the WAP architecture so that they need not concern themselves with the exact type of wireless network the application is running on. Among other capabilities, WDP provides data error correction.

Bearers

The bearers, or wireless communications networks, are at WAP's lowest level. WAP is designed to run on a variety of networks including short message services (SMS), circuit switched connections, and packet switched networks. Each type of network has pros and cons in terms of performance, delay, and errors.

A Closer Look at WAE

Hopefully, the background on the Web and WAP transaction models provide a high-level picture of this technology. Now that we've dug a bit deeper into the layers of the WAP protocol, let's pop back up and examine the place that the majority of developers spend their time. It's also the place that you should most likely concern yourself with if you're thinking of using WAP technology.

The Wireless Application Environment has four key components:

The Microbrowser. The microbrowser defines how WML and WMLScript are interpreted by a WAP-enabled device for presentation to the end user.

WML. The Wireless Markup Language is similar to HTML and defines how data should be formatted and presented to the user.

WMLScript. Similar to JavaScript, WMLScript provides some programming logic for performing calculations within an application.

Wireless Telephony Applications. WTA provides functionality so that developers can integrate microbrowser functions with the telephone. For example, an incoming call may trigger the microbrowser to search your Contact list and show the information at the time the call is received.

These elements of the WAE weren't just made up by a team of mad scientists bent on adding more acronyms to our world. WAP is based on a range of existing Internet technologies. Appendix A provides a list of books, magazine articles, and electronic documentation where additional information may be obtained. These technologies include:

- Unwired Planet's Handheld Markup Language (HDML) [HDML 1999]
- The Hypertext Mark-up Language (HTML) from the World Wide Web Consortium [Raggett 1999]
- ECMAScript, which is based on JavaScript [ECMA 1997][Flanagan 1998]
- The vCard and vCalendar exchange formats by IMC [vCalendar 1996] [vCard 1996]
- A variety of other Web technologies like URLs and HTTP [WAP 1999]

Let's look at these four components in more detail.

Microbrowser

We've already discussed what the general duties of a microbrowser are. Like a regular web browser, it submits requests for information, receives results, and interprets and displays those results on screen. There are also some secondary tasks associated with the job of microbrowser.

The microbrowser includes both WML and WMLScript interpreters. As the phone receives binary information in this format, the microbrowser interprets that data and decides how to display WML and execute WMLScript.

Though not specified in the WAP specification, the microbrowser may have additional capabilities. For example, the phone may include RAM for caching information in the same way computer hard drives cache information for regular web browsers. If so, the microbrowser will have software that helps it decide when a page should be cached, how long the information in the cache is valid, and when to remove items from the cache.

The microbrowser is also responsible for understanding the HTTP 1.1 protocol. As we've already described, the gateway is responsible for much of the translation between the WAP and HTTP protocols. However, when a request is sent from a WAP device, the microbrowser must be able to include valid HTTP information in the request so that the web server knows how to interpret the request.

Finally, the microbrowser needs to know how to manage the limited resources of the WAP device. These devices are limited in screen size, processing power, RAM, ROM, and input/output capabilities. The microbrowser is responsible for juggling the demands of this limited environment.

WML

As we've said before, the Wireless Markup Language is similar to HTML. However, WML borrows heavily from the constructs of the Extensible Markup Language (XML), the Internet successor to HTML. The creators of WML accounted for the limited resources of WAP devices. However, they kept the tag-based design of HTML and in some areas built more robust features into WML than those provided by HTML.

To see what we're talking about, look at the following snippet of HTML, which appears in web browser window in Figure 4.9:

```
<html>
 <head>
 <title>Empyrean Design Works</title>
 </head>

 <body>
 <h1>Welcome</h1>
 <p>Empyrean Design Works is a full service software design and strategy
    firm for mobile, wireless, and handheld technologies.

 </body>
</html>
```

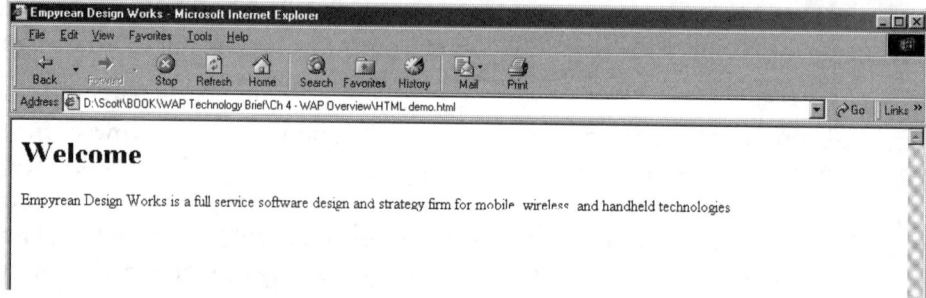

Figure 4.9 Sample HTML in a web browser.

Compare that to the following WML code sample running on a WAP phone in Figure 4.10:

```
<wml>
  <card id="first" title="Empyrean Design Works">
  <p>Empyrean Design Works is a full service software design and
     strategy firm for mobile, wireless, and handheld technologies.</p>
  </card>
</wml>
```

While the two pieces of code do not look identical, notice the similarities between HTML and WML. Instead of starting and ending the document with <html> and </html>, a WML document uses <wml> and </wml>. Also, notice that <p> is used in both languages as a way to mark a new paragraph within the document.

Both WML and HTML render similarly in their respective browsers. WAP browsers are just much more screen size challenged. We look at a sample of WML code in more depth in the next chapter. Hopefully, the example above demonstrates our point about the syntactical similarity between the two languages.

WML as a language has the following general features:

Support for text and images. This includes presentation hints like line breaks, formatting (bold, italic, and emphasis), and other placement clues. As we saw in the example at the beginning of this chapter, not all devices support all text styles. WAP-compliant devices are not required to support images (this should change over time as devices become more capable) although the protocol does support them.

Support for user input. WML includes text entry fields, choice lists, and controls that invoke tasks. For instance, you can assign a URL to a specific button on a device such that when the user presses the button, a request is sent for a new document. The WAP specification has no specific definitions on

An Overview of WAP 43

Figure 4.10 Sample WML in a microbrowser.

how user input is accomplished. For example, if a WML program includes a list of options, the user may have to make their choice by pressing hardware buttons, tapping an on-screen button, or using voice input. It's up to each device manufacturer to determine how an options list is implemented.

A variety of navigation mechanisms. Based on the Internet-standard URL naming scheme, WAP lets you move between documents. Each WAP device may also incorporate a history mechanism for documents already visited, so that the user can revisit a previous document just by pressing a Back button, much like revisiting a previous page in a web browser.

Support for multiple languages and dialects. WML provides support for multiple languages and dialects by using the 16-bit Unicode character set.

State and context management features. State management implies that variable values can be passed from document to document. Additional capabilities include variable substitution and caching of variables and documents to maximize cache hits on the device and minimize wireless server requests.

WMLScript

WMLScript adds a lightweight procedural scripting language to each WAP device. Loosely based on JavaScript, WMLScript lets programmers add intelligence to WAP programs, and reduces the necessity for requesting information from the web server.

Programmers can use WMLScript for the following:

Input Validation. As users enter data like their name, a dollar amount, or phone number, WMLScript can validate the input against some template. For example, it can check that the dollar amount entered is under $100 and includes two digits after the decimal.

User Interaction. WMLScript lets an application interact with a user without constantly needing to contact a web server for more documents. For instance, the `if...the...else` capability lets the program logic decide which document to show next or display an error message of some sort without first going over the network.

WMLScript also includes libraries that provide a wide range of functionality including math calculations, string processing, and URL manipulation, for instance.

Wireless Telephony Application Interface (WTAI)

WTAI is designed to let network operators access the telephony features of a WAP device. They can do such things as initiate phone calls using WML and WMLScript, accept incoming calls, hang up calls, send and receive text messages, and manipulate phonebook entries on the device. Besides those functions that are common to all to WAP devices, WTAI supports telephony extensions that are specific to certain wireless telephone networks like GSM and PCS.

WAP is a feature-rich application environment. It's programmability and telephony features make it very suitable for creating mobile applications. It's compact form, encryption, and error handling make WAP suitable for the challenges of wireless transactions.

WAP will undoubtedly continue to evolve to support advanced features and functionality that will appear in smart phones in the coming months and years.

In this chapter we described the various WAP layers. Fortunately, you can, for the most part, ignore the details of those layers. Trust them to work properly. If you're compelled to delve into more details, you should download the WAP specifications from the WAP Forum's web site at www.wapforum.com.

In the next chapter we focus on the layer of most interest to you, the WAP Application Layer (WAE).

CHAPTER 5

The WAP Application Environment

In Chapter 4 we described each of WAP's layers and its duties. We also discussed how data from a WAP application is transformed into a format suitable for transfer across a wireless network.

For our purposes, the most important WAP layer is the WAP Application Environment (WAE). In enterprise and WAP service development, the development teams spend most of their time here. In addition, WAP users spend most of their time interacting with a WAP device's microbrowser, which is part of the WAE.

In this chapter, we explore WML and WMLScript, the programming languages used to create WAP applications. Don't worry, this chapter isn't too technical. It does, however show the capabilities and challenges of using WAP. If you are interested in more details on WML and WMLScript, take a look at *Programming Applications with the Wireless Application Protocol* (Mann 1999).

The Microbrowser

A microbrowser, which is similar in function to a traditional web browser, is built into all WAP devices. In most devices the microbrowser uses a two- to five-line display on a smart phone. In other devices, such as a Palm connected organizer or Ericsson's MC218, the microbrowser has a much larger screen, which lets it display significantly more data. The microbrowser's main function is to interpret the WML and WMLScript documents retrieved from web servers.

Some microbrowsers can display both text and graphics, which is handy when screen real estate is at a premium. For example, showing an icon representing a partly cloudy day takes up less space than writing "Partly Cloudy" on the screen. Today's WAP devices have a mix of text-only and text/graphics displays. Manufacturers are, in fact, moving to highly graphical displays able to display several shades of gray, and in the not too distant future, color.

WML

WAP devices retrieve data from the Internet. The built-in microbrowser retrieves documents written in the Wireless Markup Language (WML). WML is designed for creating applications that run on WAP-enabled devices. WML looks similar to HTML, though it borrows heavily from the Extensible Markup Language (XML)[Bray 1998]. XML is considered the successor to HTML [Ragget 1999]. It offers many robust features that are not available within HTML.

Some of the basics of WML include elements, attributes, and the card/deck metaphor. While we don't explore WML programming in any depth, it is worth looking at the basic pieces of a WML document to gain an appreciation for the abilities of the language.

Elements and Attributes

A WML document, at its most basic level, is constructed from a set of *elements*. An element consists of a *beginning tag* and an *ending tag* with usually displayable content in between. The concept of an element is not exclusive to WML. It is also part of HTML and XML.

Let's look at an example. The following

```
<p>
```

is a beginning tag. It tells the microbrowser to start a new paragraph. Everything before the ending tag

```
</p>
```

is treated as content and included in the paragraph. The two tags together constitute an element.

Here's a more detailed example. This paragraph element contains content for display on the screen.

```
<p>
   This is a new paragraph.
</p>
```

The text between the tags begins on a new line on the display.

A tag may contain *attributes*. Attributes define additional qualities of the tag. Building on our previous example, here's another paragraph.

```
<p align="right">
   This is a new paragraph.
</p>
```

`<p>` and `</p>` define the paragraph element. We've added the `align="right"` attribute to the beginning tag. This tells the microbrowser that the content should be right justified. If the microbrowser supports text alignment, the text will be right justified.

Many WML attributes that are assigned to individual elements are optional, meaning that according to the WAP specifications, microbrowsers do not have to support them. They are hints or suggestions as to how the content should be displayed, not rules. Consequently, as with other desktop web browsers, content that appears one way on one WAP device may appear differently on another.

Many WML attributes have default values if you do not specify one. In the previous example, we don't have to tell the microbrowser how to justify the paragraph text. It defaults to "left" if you do not explicitly tell it.

Decks and Cards

In the world of HTML, a web browser requests a document from a web server. If the request is for an HTML page and the server returns it, the web browser renders the page on-screen. Frequently, the document occupies more than the entire screen, forcing the user to scroll up and down, and in some cases left to right.

Due to the limitations of screen size on most WAP-enabled devices, the creators of the WAP specification opted to take a different approach than retrieving and displaying an entire document. Instead, they call a document retrieved from a server a *deck*. A deck, the smallest piece of data that can be retrieved from a server, is made of several cards, as shown in Figure 5.1. Conceptually, a *card* is a screen of data that a user reviews or uses to make a selection or enter requested information.

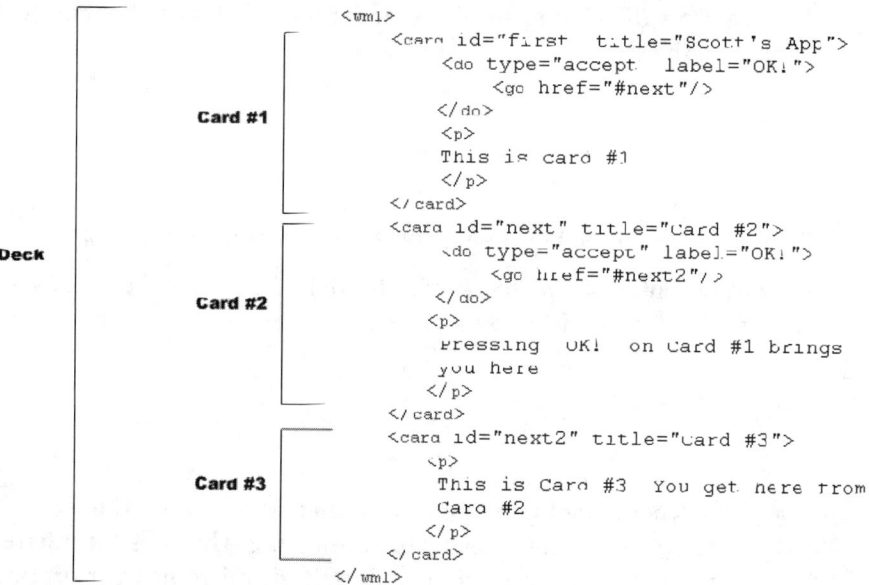

Figure 5.1 A deck with three cards.

More generically, a complete WML deck is structured like this.

```
<wml>
 <head>
 .
 . head information
 .
 </head>
 <template>
 .
 . template information
 .
 </template>
 <card>
 .
 . card information
 .
 </card>
 .
 . additional optional cards
 .
</wml>
```

The <head></head> and <template></template> elements, which we don't describe here, are optional. At minimum, a deck must contain at least one card to be valid.

WML Features

We've looked at the basic syntax of elements and attributes and cards and decks. Now, let's go through some of WML's features.

WML elements can be divided into three groups. The first is content, which relates to the information presented on screen and how it is formatted for the user. The second group has to do with tasks and events. When the user presses a button, he or she creates a task that, in turn, triggers an event that causes some action to happen. Finally, WML has data entry features to handle a variety of input methods.

Content

The information shown on the screen of a WAP device is content. Content can be unformatted and formatted text as well as graphics on some devices. Not all devices support all WAP content elements like italicized text and graphics. Instead, the elements are treated as hints or preferences as to how you'd like the content to appear on-screen. They don't necessarily guarantee how it *will* appear. While web designers prefer uniformity of presentation across web browsers, WAP does not guarantee uniformity because of the variety of devices that exist.

Whether this is considered a feature or a drawback depends on who you are. Manufacturers like it because it gives them flexibility in designing limited resource devices. Application designers and content providers are less accepting because they cannot guarantee that two users on two different devices will have the same experience.

With these caveats in mind, WAP content falls into the following categories:

Aligned Text. There are attributes to left, center, and right justify text and an element to introduce line breaks. Figure 5.2 shows the following code sample, demonstrating different types of aligned text:

```
<wml>
 <card>
  <p align="center" mode="wrap">
   Centered Text </p>
  <p align="right" mode="wrap">
   Right Text </p>
 </card>
</wml>
```

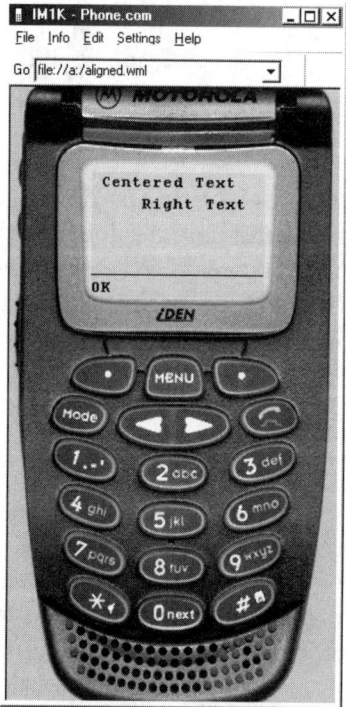

Figure 5.2 Aligned text.

Styled Text. Developers can identify text as bold, large, emphasized, italicized, small, strong emphasized, and underlined. Figure 5.3 shows the sample below. Note the `<br>` tag for starting new lines:

```
<wml>
 <card>
  <p>
   <i>Italic </i> <br/>
   <small>Small </small> <br/>
   <big>Big </big> <br/>
   <u>Underline </u> <br/>
   <b>Bold </i> <br/>
   <em>Emphasize </em> <br/>
  </p>
 </card>
</wml>
```

Tables. While visions of Excel may dance in your head, WAP tables are for elementary screen layout only. While not powerful, they make lining up columns of data a breeze. Figure 5.4 shows how easy a flight schedule is to read using a table.

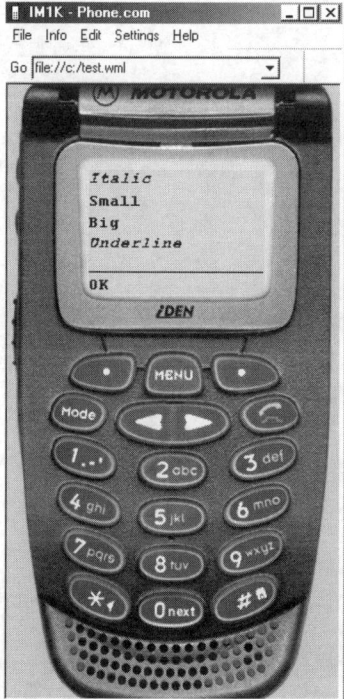

Figure 5.3 Styled text.

Images. There may be times when a small graphic saves a large amount of precious screen space yet communicates as effectively as a lot of text. The WAP forum has defined its own graphics file format, the Wireless Bitmap (WBMP). Think of it as the wireless version of JPEG or GIF, the two most common graphics file formats associated with the Internet and web browsing. All WAP gateways, the intermediates between WAP clients and Internet sites, are required to recognize the WBMP format. Some can also convert other graphic formats to WBMP. Figure 5.5 shows a WML card containing a small graphic.

Tasks and Events

So far, we've discussed only the presentation and layout features of WML. There are also WML elements—tasks and events—for handling instructions based on certain conditions. Tasks are simply things that a WML program does in response to the user triggering a certain event. As you'll see, tasks and events are a powerful combination that allows for a wide range of navigation and decision options for applications.

Figure 5.4 A table example.

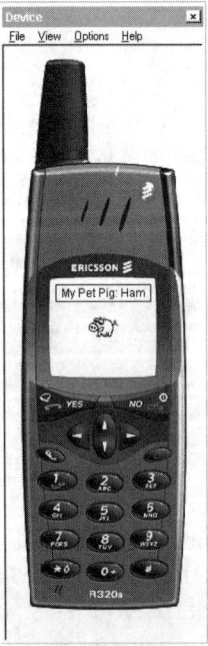

Figure 5.5 A Wireless Bitmap example.

When a user presses a button on a WAP device, an event occurs. A WML program can recognize this event and initiate a task. The task may cause a different card in a deck to appear on screen, load an entirely different deck, or refresh the contents of the display.

The following card looks like Figure 5.6 on a WAP device. This first line of the card gives the card a title, "URL Demo." The second through fourth lines define a task and event. If the "accept" event occurs, the WAP device executes a task and requests the document "sample.wml" from a server.

```
<card id="LoadURL" title="URL Demo">
  <do type="accept" label="Load URL">
    <go href="sample.wml"/>
  </do>
  <p>Press 'Load URL' to retrieve the next deck.</p>
</card>
```

Which button generates an "accept" event? That depends on the device. WAP does not define a specific button for "accept" events. Instead, it just mandates that every WAP device must have some sort of method available to the user that generates "accept" events. In Figure 5.6, for example, the key to the left of the Menu button generates an "accept" event.

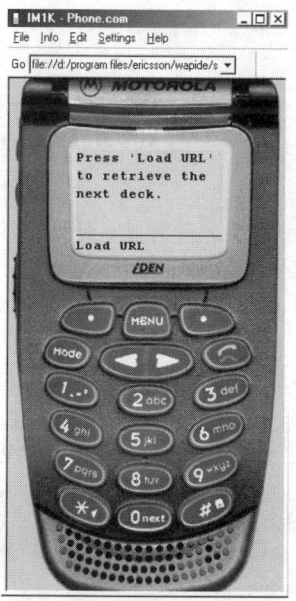

Figure 5.6 A task and event demo.

WML supports four types of tasks: `<noop>`, `<prev>`, `<refresh>`, and `<go>`. `<noop>` and `<refresh>` are the less interesting of the four tasks. The former does nothing, as its name implies. `<refresh>` updates the user's display.

`<prev>` is used to navigate through a microbrowser's history. It retrieves previously visited URLs. Its functionality is not unlike pressing the Back button in a desktop web browser. The number of URLs you can go back is dependent on the individual capabilities of the phone. WAP does not specify how the `<prev>` task is implemented or how many URLs it can remember. Rather, it only specifies what should occur if a developer uses the task.

`<go>` is straightforward. Depending on its attributes, `<go>` gets a web page from a server or posts user data to a server. This task is particularly useful because it lets an application, for example, retrieve a particular URL if the user presses a button, taps the screen, or uses voice input to make a choice.

Events come in three flavors: `<anchor>`, `<onevent>`, and `<do>`. `<anchor>` elements work in the same fashion as anchors within HTML. For example, you've used an anchor in a regular web browser if you've clicked a link on any web page. The link may move you to a different spot within the same page or it might load a completely different page.

WML anchors do the same thing, but they move the user between cards within the same deck or retrieve a different deck. Specifically, if you use an anchor you must specify a task to perform for the event. The available tasks you can assign to an anchor are `<prev>`, `<refresh>`, and `<go>`.

An `<onevent>` event fires when one of four things happens:

- The user navigates to a new page.
- The user navigates to a new page as a result of using the microbrowser's history through the `<prev>` task.
- A program-defined timer expires.
- The user selects an item from a pick list within an application.

When an `<onevent>` event occurs, any of the four tasks listed above may occur.

Finally, WML supports `<do>` events. Every WAP device has a set of predefined user interface widgets that are available to a user. A widget could be a real button on a smart phone, a voice command, or a touch screen. How the WAP device manufacturers support these widgets is up to them. However, they must define a way for supporting the following seven types of `<do>` events:

accept. A positive acknowledgment.

prev. Navigate backward through the history of URLs.

help. A help request from a user.

reset. Reset the device's context.

options. A context-sensitive request for options.

delete. Delete the current item, element, or choice.

unknown. A generic event.

To cite a real-world example, let's look at the help event. WAP device manufacturers must support a facility for letting users request help within an application. Pressing a button on the device, speaking the word "help," or tapping a button on the interface are all ways that event can be implemented. It's up to the manufacturer to pick the most appropriate mechanism.

Data Entry

Applications are more useful when the user can provide input. Data input may come in a variety of methods including picking from a list or using a keypad. Unlike your home computer, which has a full-size keyboard and mouse, WAP devices are typically limited in their input methods. WAP does not dictate how input should occur; it leaves that up to the device manufacturer.

The two most typical input methods are the telephone's numeric keypad and additional device buttons. With the keypad, a user must press the number "7" key four times to type an "S" and eight times to type an "s." This method is very inefficient. Additional buttons facilitate moving between on-screen choices and letting the user make selections. Many companies are working on alternate input methods. Before we explore these technologies, let's look at WML's input elements.

The `<input>` element supports the entering of a string of characters. How those characters get entered is device dependent. It can be by selecting a letter from the keypad or using a stylus with handwriting recognition directly on the screen, for instance.

There are a number of optional attributes for controlling the format of the input. Here are the most important:

emptyok. This attribute let's the user enter nothing for a particular field.

format. Developers can create input mask strings that force the user to enter the data a certain way. For example, you can make the user enter a complete nine-digit zip code using the format `"xxxxx-xxxx"`.

maxlength. This attribute specifies the maximum number of character a user can enter for an input field.

type. The values for this attribute can be `"text"` or `"password"`. If `"password"` is set, the user's input is obscured so that it is unreadable.

On a desktop computer, this is normally done by showing asterisks instead of the actual input characters.

value. This attribute lets a programmer define a default value for a field if the user does not input anything.

The other major input method uses the <select> element in conjunction with <option> and <optgroup> elements. Figure 5.7 shows the <select> element in action. It lets a developer create a set of choices from which the user can select an item. Optionally, the <optgroup> element creates a hierarchical set of options. For example, a question might be, "What's your favorite hobby?" The choices could be sports, reading, and social activities. Using an <optgroup> element, the user can select sports from the list and then drill down to a list of specific sports.

Input Alternatives

We mentioned previously that companies are working on alternative input methods for all types of mobile devices. Each of the following has advantages and disadvantages. Here are a few of the most promising methods.

Voice Recognition. Voice recognition has the potential to speed data entry. Initially, it would let a user choose a selection from an option group by saying the number associated with the option. More advanced voice recogni-

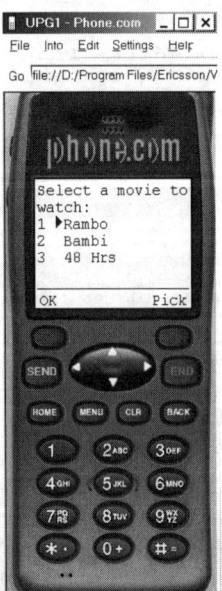

Figure 5.7 A group of choices as an input method.

tion would let a user say a co-worker's name and then the command "Dial," or simply speak into the microphone as freely as he would to a friend. However, it will be some time before we get to that level of recognition. Issues include the processing power needed to handle free-form speech and understanding user's accents.

Predictive Input. This technology is currently finding its way into smart phones. By typing on a numeric keypad, the program attempts to guess the word you're entering. The sequence "8-4-3" can be interpreted as both "The" and "Vif." Using algorithms for identifying commonly used words and the context of the application, the program would assume you mean "The." The limitation is simply that the software may guess incorrectly.

Handwriting Recognition. The Apple Newton PDA was one of the first devices to extensively use handwriting recognition. The Palm connected organizer uses a related technology for input, but forces the user to learn its method for writing letters and numbers. The advantages are obvious. The challenges are that it may take too much processing power, the software may interpret handwriting incorrectly due to sloppiness, and WAP devices become more expensive to manufacture when you include touch-sensitive screens.

We've emphasized the fact that WAP does not dictate how input is done. It simply gives a framework for the type of input it accepts. None of the preceding input mechanisms will single-handedly solve the input problem. Each has pros and cons. It will be some time before a set of best practices emerges from this industry.

WMLScript

When the Web first became popular and HTML was the only technology for creating pages, many people noticed its shortcomings. One of the most evident was the lack of techniques for adding client-side logic to pages.

For example, if a web page contains a form that asks for your age, the browser has no way of making sure you enter a valid number. You could enter a negative number or an address instead. Once you submit the form to the server, the server can respond with an error message if you've entered an invalid value. This type of transaction is a waste of time and data bandwidth. What's needed is a way of putting programming logic into the web page itself to check for things like illegal values.

The creators of WAP designed their own client-side programming language, WMLScript, for this purpose. WMLScript is a derivative of ECMAScript [ECMA 1997], which is itself derived from JavaScript [Flanagan 1998] and Self

[Ungarand and Smith 1987, pp. 227–241]. WMLScript is a lightweight programming language used to add functionality to WML. The WAP Forum made WMLScript as minimalist as possible so that it's easy to learn, uses little microbrowser memory, and creates small executables that can be quickly transmitted in a wireless environment.

Some of WMLScript's key features include:

Data Types and Variables. Programmers can declare variables for use in their scripts. The variables are loosely typed, meaning they change their type depending on the value last assigned to them. WMLScript recognizes the following types: Boolean (true/false), integers, floating-point (decimal) numbers, and strings.

Statements. Within functions, developers can use standard programming and flow control statements. These include `if...then...else` decisions, `while` and `for` loops, and assignment statements with calculations.

Library Functions. To add functionality to basic WMLScript, the creators added some standard function libraries for performing a variety of tasks. These include: Lang, Float, String, URL, WMLBrowser, and Dialogs.

While it's outside the scope of this book to discuss all of these functions, it is worth noting what is contained in a couple of the WMLScript libraries. The Lang library contains arithmetic, conversion, environment, and random number functions. The Float library has the equivalent environment and arithmetic functions for manipulating floating point numbers, numbers containing a decimal point. Finally, the String library has functions for manipulating strings. You can compare two strings, create sub-strings from a larger one, and format strings to display a certain way on screen.

Combined with WML, WMLScript makes it possible for programmers to create robust programs suitable for almost any data processing function from games to enterprise applications. If you're curious about how to actually create a WML/WMLScript application, see "Resources" at the end of this book for pointers to Software Development Kits (SDKs). We also mention several of the most important SDKs at the beginning of Chapter 4, An Overview of WAP.

CHAPTER 6

WAP Client Software, Hardware, and Web Sites

Products, services, and standards without third-party support don't last more than a few months. The same is true for WAP.

In this chapter we give you an overview of WAP microbrowsers. They come in two flavors. The first, called OEM (Original Equipment Manufacturer) microbrowsers, is software that a cellular telephone manufacturer can license for inclusion in a WAP device. The second type of microbrowser includes those that individual consumers can download and use on a handheld or desktop computer. Some companies, such as AU-System, make both consumer and OEM versions of their browser.

We also include an overview of some of the WAP-compatible devices currently available. Most of them were introduced in the six months prior to May, 2000. The cellular telephone business is a consumer electronics market. In the next twelve months, most of the current models will probably be replaced by newer ones. This list is just a representative cross-section to give you an idea of the types of devices currently being manufactured.

OEM Microbrowsers

Several companies have developed WAP microbrowsers for inclusion in WAP devices that they or other companies manufacture. Here are the major offerings.

UP.Browser

Long before WAP ever existed, Phone.com (www.phone.com) negotiated deals with numerous telephone manufacturers to use its UP.Browser. This browser interpreted WAP's predecessor, HDML. In fact, HDML is still the most prevalent technology for delivering Internet content to cellular telephones in the United States. This is quickly changing, however, as providers migrate to WAP and the latest incarnation of the UP.Browser, which does interpret WML and WMLScript.

Over 20 wireless telephone manufacturers license the UP.Browser for their WAP-compatible devices. Current licensees include Hitachi, Sanyo, Sony, Motorola, QUALCOMM, and NeoPoint. Since Phone.com assisted in the creation of the original WAP specification, expect them to remain a dominant player in the WAP market.

Ericsson WAP Browser

Ericsson (www.ericsson.com/wap) is another major player in the WAP market with excellent development support for their devices. They have their microbrowsers currently built-in to three smart devices: the MC218 mobile companion and the RS320 and R380 telephones. As we describe in the next section of this chapter, each of these devices has drastically different designs with different screen sizes and computing abilities.

Ericsson also offers a version of AU-System's WAP browser for Palm Computing connected organizers. Ericsson's version is modified to run Ericsson's implementation of the WAP standard. This microbrowser is available for download from Ericsson's web site in their development area (www.ericsson.com/developerszone).

Mobile Explorer

Never content to allow a competitor to enter a market unencumbered, Microsoft has created its answer to WAP: Mobile Explorer (www.microsoft.com/wireless). Mobile Explorer is the first microbrowser to do double duty. It can interpret and display both HTML and WAP 1.1 content.

Microsoft's research estimates that over 525 million cellular telephones will be sold worldwide in 2003. Over 90 percent of those devices are expected to include microbrowsers. Microsoft intends to license Mobile Explorer as a microbrowser for all types of devices to capture a significant share of the WAP market.

Figure 6.1 The AU-System WAP browser for the Palm OS.

AU-System

AU-System (www.wapguide.com) is a Swedish company that has a microbrowser architecture they port to different devices. Their WAP browser, shown in Figure 6.1, comes in versions for Windows CE, EPOC, the Palm OS, and the REX platform, among others. AU-System licenses this browser to others for use in their devices.

AU-System clients include Compaq, Ericsson, LGIC, SK Telecom, and Samsung Electronics. Ericsson, for example, adapted the AU-System WAP browser to run Ericsson's WAP implementation on the Palm OS. The generic version of this browser for the Palm OS is available from AU-System's web site. Connect a modem to a Palm handheld and you're off and running.

Consumer Microbrowsers

You don't necessarily need a cellular telephone to view web sites that host WML content. A handheld or desktop computer and WAP browser work together quite well. There are several products available at no charge on the Internet. Additionally, many of the microbrowser manufacturers, including Phone.com, Nokia, and Ericsson, make emulators available as part of their WAP Software Development Kits (SDKs) so that developers can simulate

WAP devices on a PC. We include URLs for downloading these SDKs in Chapter 4, An Overview of WAP.

WAPMan

WAPMan was created by The Edge Consultants in Singapore (www.wap.com.sg). They make two WAP clients. The first client, shown in Figure 6.2, runs within Windows 95, 98, and NT. The other, shown in Figure 6.3, operates on Palm, Inc. connected organizers. Both are WAP 1.1 compliant browsers and can connect to WML sources through WAPgate, the Edge Consultants' WAP gateway. Using an Internet-connected PC or Palm, you can view WML-encoded web sites.

When you enter the URL of a WAP-compatible document, WAPMan retrieves the requested page and displays it. All requests are routed through their WAPgate product, then out to the web site containing the WAP document you're interested in. WAPMan is a great way to get your feet wet with WAP, especially if a cellular phone-based WAP service is not currently offered in your area. It simulates the experience of using a real smart phone. The interface has a number of additional buttons on it including Homepage, Back, Stop, Go, and Reload buttons. You can also bookmark your favorite WML sites.

One of the disadvantages of the current line of WAP-compatible cellular telephones is their limited screen real estate. On a Palm connected organizer, the WAPMan client can display more information if it's available. You interact with the client by tapping links on the screen instead of the usual smart phone device buttons. Because of the variation in input methods and dis-

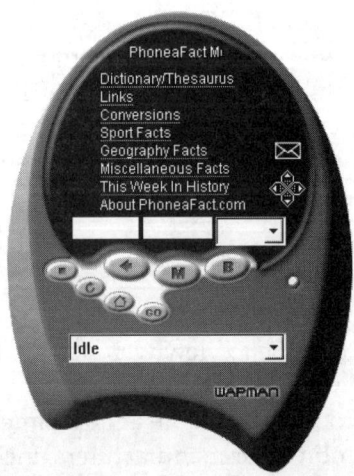

Figure 6.2 WAPMan running on Windows 98.

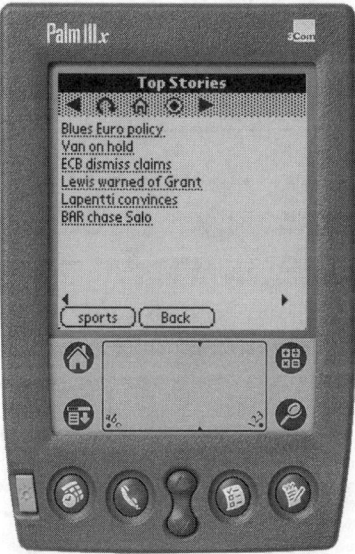

Figure 6.3 WAPMan running on the Palm OS.

plays, the Palm version of WAPMan makes use of the traditional web metaphor by including navigational buttons at the top of the screen.

WinWAP

From Slob-Trot software in Finland (www.slobtrot.com) comes WinWAP, a full-size WAP browser for the Windows platform. It runs under Windows 95, 98, and NT and lets you view the growing number of WML and WMLScript pages on the web. Shown in Figure 6.4, WinWAP is remarkably similar to Internet Explorer and Netscape Communicator in both form and function. Because of this, WinWAP is one of easiest ways to start experiencing WAP content.

Slob-Trot has two versions of their browser: Light and Pro. WinWAP Light reads the information directly from a web server and does not use a WAP gateway to encode and encrypt the information. For most purposes this works well, since performance through a PC connected to the web is almost guaranteed to be faster than a wireless WAP connection. The Pro version of the software adds a feature in which you can specify a gateway so that all communications are translated, encoded, and encrypted, though they never go through the air.

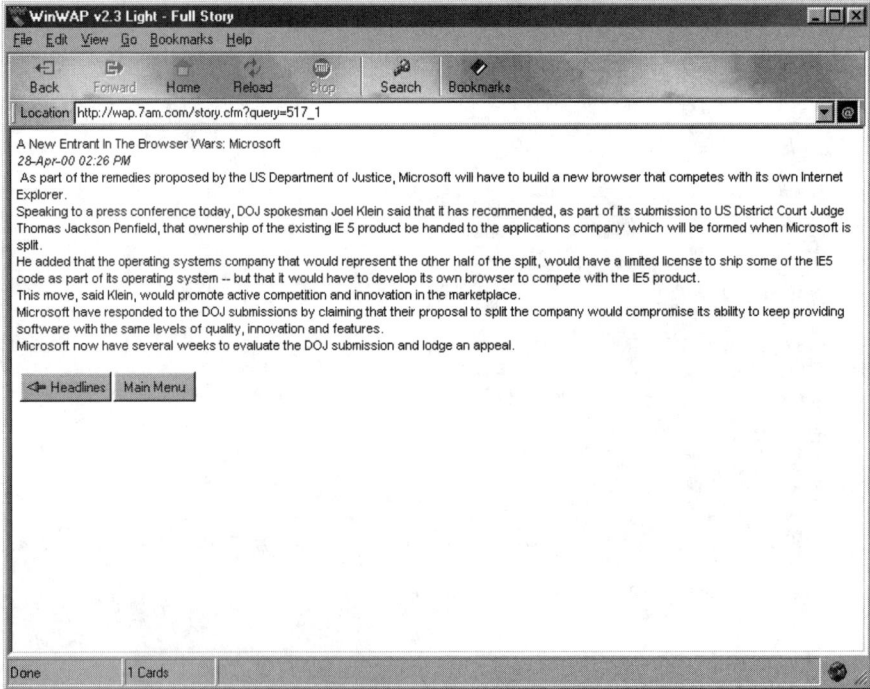

Figure 6.4 WinWAP running on Windows 98.

4thpass KBrowser

4thpass (www.4thpass.com) offers a Java-based cross-platform WAP browser based on Sun Microsystem's Java KVM. Java KVM is a small memory footprint Java implementation optimized to run on devices with limited resources. In theory, one version of KBrowser works on any platform that supports the Java KVM. Currently, the only commercially available devices that the KVM runs on are Palm, Inc.'s organizers.

WAP Devices

There are hundreds of cellular telephone models on the global market. Some are analog, some are digital, some are both. Some can transfer data while others are limited to voice. It can be confusing to determine which telephones support the WAP standard.

This list contains a subset of the available WAP-enabled cellular telephones. Other manufacturers, including Samsung, Kyocera, and Sony, make WAP mobile devices. It will be a few years before all cellular phones support

WAP, but as you can see, the companies listed here represent a manufacturer's Who's Who.

Nokia 6210/6250

The Nokia (www.nokia.com) 6210 (see Figure 6.5) and 6250 are similar telephones. Their primary difference is that the 6250 is ruggedized for wet, dusty, dirty, extreme environments. Both devices are capable of operating at speeds of up to 43.2 Kbps using HSCSD (High Speed Circuit Switched Data) for use as a high-speed wireless data modem.

Product: Nokia 6210

Networks: GSM 900, GSM 1800, dual-band GSM 900/1800

Availability: Europe, Africa, Asia Pacific

Talk Time: 2.5–4.5 hours

Standby Time: 55–260 hours

Display: Full graphical, 96 × 60 pixels

WAP: Nokia's WAP 1.1 microbrowser

Other features: Voice dialing, text and picture messaging, built-in calendar software, PC synchronization

Figure 6.5   The Nokia 6210 WAP phone.

Nokia 7110

The Nokia 7110, shown in Figure 6.6, is this company's start-of-the-art WAP telephone. It boasts a variety of advanced features including three fonts, multi-lingual versions, including support for Chinese character input, and GSM data transmission capabilities at up to 14.4 Kbps.

Product: Nokia 7110

Networks: Dual-band GSM 900/1800

Availability: Europe, Africa, Asia Pacific

Talk Time: 1.75–7 hours, depending on the battery

Standby Time: 35–430 hours, depending on the battery

Display: Full graphical, 96 × 65 pixels

WAP: Nokia's WAP 1.1 microbrowser

Other features: Call management, predictive input, calendar, phone book, games

Figure 6.6 The Nokia 7110 WAP phone.

Motorola

Motorola (www.motorola.com) has several families of WAP-enabled smart phones, divided into two groups.

The first group uses TDMA-based iDEN technology. These telephones include the Motorola i500 Plus, i700 Plus, and i1000 Plus. They typically include vibrating call alert, a high-quality, built-in speakerphone, two-way-radio, paging, and digital data capabilities. Figure 6.7 shows the Motorola i1000 plus.

The second group of products includes the TimePort, TalkAbout, and StarTAC Telephone. There are at least seven models that come in a variety of sizes, shapes, and feature sets for use in different parts of the world. Some are pure digital devices, others are hybrid analog/digital. They support all major networks, including GSM, TDMA, and CDMA. Check Motorola's web site at www.motorola.com for details on the current models.

Product: Motorola i1000 plus

Networks: iDEN

Availability: North America

Talk Time: 3 hours

Standby Time: 1 hour

Display: Full graphical, five lines

WAP: Phone.com's UP.Browser

Other features: Speakerphone, two-way radio, paging, digital data modem

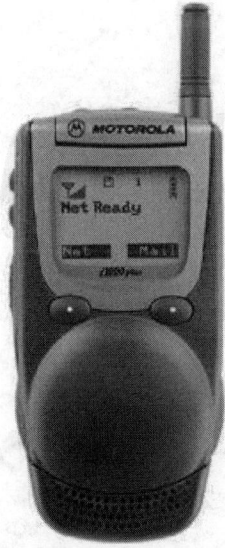

Figure 6.7 The Motorola i1000 plus.

Ericsson R320/R380/MC218

Ericsson's (www.ericsson.com) WAP-compatible devices demonstrate the flexibility of the WAP standard to handle a variety of device display sizes. The R320, shown in Figure 6.8, has a limited display. It can display 5 lines of text or 101 by 65 pixels of information.

The R380, shown in Figure 6.9, has a larger, more feature-rich display. It can show cards of up to 360 by 120 pixels. It uses the EPOC operating system to essentially combine a dual-band cellular telephone and a feature-rich hand-held operating system. The R380 is not yet available.

Finally, the MC218 mobile companion, shown in Figure 6.10, has a half-VGA screen able to display 640 by 240 pixels. While it does not have a cellular phone built-in, it can communicate to the Internet through its infrared port to an Ericsson mobile phone.

Product: Ericsson R320

Networks: Dual-band GSM 900/1800

Availability: Europe, Asia Pacific

Talk Time: 4.5 hours

Standby Time: 100 hours

Display: Full graphical, 96 × 65 pixels

Figure 6.8 The Ericsson R320s WAP phone.

WAP Client Software, Hardware, and Web Sites 69

Figure 6.9 The Ericsson R380 WAP phone.

WAP: Version 1.1 microbrowser

Other features: Call management, voice input, infrared, calendar, contact, and games software, PC synchronization

Product: Ericsson R380

Networks: Dual-band GSM 900/1800

Availability: Europe, Asia Pacific

Talk Time: NA

Standby Time: NA

Display: Full graphical, 360 × 120 pixels

WAP: Ericsson version 1.1 microbrowser

Other features: EPOC operating system with extensive PDA functionality and PC synchronization

Product: Ericsson MC218

Networks: Works with IR-enabled Ericsson telephones

Display: Full graphical, 360 × 120 pixels

WAP: Ericsson's version 1.1 microbrowser

Other features: EPOC operating system with extensive PDA functionality and PC synchronization

Figure 6.10 The Ericsson MC218 WAP-enabled PDA.

MobileAccess T250

The MobileAccess T250, shown in Figure 6.11, is manufactured by Mitsubishi (www.mobileaccess.com). It is a dual-band, dual-mode TDMA/DAMPS phone with a built-in CDPD data modem. TDMA is used for voice, CDPD for circuit switched data.

Product: Mitsubishi T250

Networks: 800/1900 MHz TDMA, 800 MHz AMPS

Availability: United States

Talk Time: 2–3 hours, depending on mode

Standby Time: 15–280 hours, depending on mode

Display: 10 line × 16 character bit-mapped display

WAP: Phone.com's UP.Browser version 3.1

Other features: SMS (Short Message Service), PC synchronization

Figure 6.11 The Mitsubishi T250 MobileAccess WAP phone.

NeoPoint 1000/1600

One of the more innovative phones out today, the NeoPoint (www.neopoint.com) 1000 shown in Figure 6.12, incorporates Phone.com's UP.Browser for viewing WAP sites on its unusually large 11 line screen. The NeoPoint 1600 pulls double duty as a digital and analog cellular telephone for increased coverage.

Product: NeoPoint 1000/1600

Networks: CDMA (1000), 1600 CDMA/AMPS (1600)

Availability: United States

Talk Time: 3 hours

Standby Time: 30–40 hours

Display: Full graphic, 11 lines

WAP: Phone.com's UP.Browser version 3.1

Other features: Voice activation, PC synchronization, calendar, address book, call management, predictive input, text messaging, e-mail

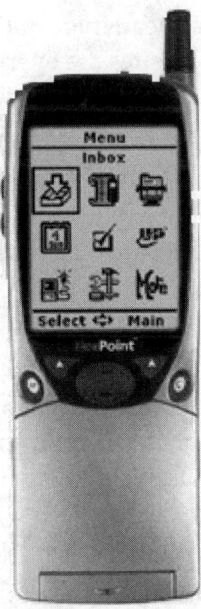

Figure 6.12 The NeoPoint WAP phone.

Consumer WAP Sites

Many herald the coming of WAP as the easiest way to retrieve corporate and sales information. It's easy to understand why businesses are so excited about this technology. Utilizing the existing and future cellular telephone infrastructures, mobile workers can exchange business information. However, consumer use of WAP will dwarf business users simply because of the number of people in the world.

The convenience, cost effectiveness, and pervasiveness of smart phone technology will make it a primary means of retrieving Internet information. For those already technology savvy, using a newspaper to find a movie time is archaic. A high-speed Internet user can more easily look up the times, read a review, and watch a trailer at his or her convenience. In fact, if that person's computer is on all of the time, it becomes the preferred way of retrieving information due to its completeness and added value.

The next logical step is the ability to access such information on the road. Knowing which cellular antennae a smart phone user is connected to opens them to a new world of possible services. You can tell the user where the nearest grocery store, gas station, movie theater, or coffee shop is located. That information might also include driving directions to the establishment. Retailers can advertise their services to travelers within the area and information can be tailored to the local market as people pass through.

While many of these wonderful services have yet to appear, informational services abound. News, weather, sports, and horoscope sites are everywhere. The number of sites is in the hundreds (April 2000) and will undoubtedly climb into the thousands by the end of the year.

Most cellular telephones provided by carriers come pre-programmed to connect to a gateway and a set of services provided by the carrier. You're also allowed to enter the URL of any other WAP site if you know the address. Fortunately, a microbrowser lets you bookmark these sites in the same way a regular web browser does. Once you've defined the sites you like, your WAP experience becomes richer and more expedient.

Here are a few of the more interesting consumer sites currently available. Note that some of the sites have PC-based web components that supplement the WAP portions. For example, you should visit 123jump.com's site to create a stock portfolio before attempting to view it from the WAP device. If, however, you're looking for a great place to start locating sites with WAP content, try one of the search engines listed below.

WAP Search Engines (PC and WAP) www.fast.no, www.wapaw.com, www.gelon.net, www.waply.com. These four sites represent both traditional search engines that build their lists by scouring the Internet for WAP sites, and directories to which people submit URLs. The first two sites may be accessed through either a web browser or a WAP microbrowser. The last two can only be used with a PC's web browser.

WAPdrive (PC and WAP) www.wapdrive.net. This clever service lets members create their own WAP pages. Once you register, you're taken to your homepage on which you can add a logo, create WAP content, and create links from your homepage to other sites. While it's a bit rough around the edges and missing some of its advertised services like e-mail and calendaring, the potential is there for this site to be the MyYahoo! of the WAP World. To get at its more advanced features, be prepared to write your own WML cards and decks.

PhoneAFact.com (WAP) www.phoneafact.com/index.wml. This site has both fun and useful services. From the homepage you can use a dictionary and thesaurus, traverse links to known news sites, get conversion information, learn miscellaneous, sports and geographic facts, and peruse what occurred today in history.

Yahoo! (WAP) wap.yahoo.co.uk, wap.yahoo.de. Two versions of Yahoo!'s world famous service are available for WAP. The first is for the UK and Ireland, the second for Germany. Yahoo! includes an extensive set of features for WAP devices including a customized MyYahoo! page, e-mail, calendar, address, financial information, news, weather, and sports. Links from these sites take you to Yahoo! sites in France and the United States. Yahoo! has also been hard at work licensing its WAP services to carriers. For example, purchasers of Sprint PCS phones have access to Yahoo! services through a portal that Sprint PCS makes available to its subscribers.

Excite (WAP) www.excite.co.uk/wap. Not to be outdone by its competitor, Excite offers WAP services with the standard news, weather, entertainment, sports, and money information.

123Jump (PC and WAP) www.123jump.com/wap. 123Jump dedicates itself to financial information. The information is detailed and extensive both on the web site and through its WAP service. Get regular and detailed stock quotes, read news, view the largest gainers and losers, and peruse your portfolio. Use the regular web site version, located at www.123jump.com, to customize your stock portfolio.

BBC News (WAP) www.bbc.co.uk/mobile/mainmenu.wml. One of the world's most respected news providers has a WAP service. Read news, sports, weather, television listings, sci-tech data, and travel information wirelessly.

Orktopas (WAP) www.orktopas.com/LnH/Translate.wml. Orktopas is another portal with links to a variety of WAP sites. One link in particular is noteworthy. ITranslator lets you type in up to 50-character phrases and translate them between six different languages: English, German, Spanish, French, Italian, and Portuguese. This is handy when traveling internationally.

WAP88.com (PC and WAP) www.wap88.com. Instant messaging is a heavily used application all over the world. WAP88.com brings it to WAP telephones leveraging the ICQ standard. Trade messages with any ICQ users from your WAP phone. WAP88.com offers other services including e-mail and games.

These are but a few of the hundreds of worldwide WAP services. To make the most of your WAP experience, use one of the four WAP search engines and a program like WinWAP Light to screen your WAP sites ahead of time. Given the limited input capabilities of the current crop of WAP-enabled telephones, this approach will save you time and effort and let you bookmark only those sites of interest.

The Internet and everything surrounding it change very quickly. By the time you read this book, chances are good that the devices, software and web sites will have changed. Think of this chapter as a historical snapshot, not a definitive reference.

CHAPTER 7

WAP Gateways

Most of the focus of this book is on WAP clients. However, a lot goes on behind the scenes at the WAP gateway. This piece of software is responsible for retrieving information from the Internet, securing transactions, and translating content from WML and WMLScript to a binary-encoded format suitable for wireless transport. Enterprise IT managers will have great interest in these aspects of implementing a WAP solution.

A Note on Terminology

Throughout this book we interchangeably use the terms server, gateway, and proxy server to identify the software that resides logically between WAP clients and Internet-based content servers. We should clarify their differences.

Server is a generic term used to describe virtually any computer on the Internet that is capable of responding to document requests either from browsers or other servers. When you enter a URL for a particular web site, you are requesting a document. (If you don't specify a document name, the web server at that site has a default document it sends you.) If that server sends that document back to you so that you can look at it on your browser, it is *serving* you the document. Hence the term server.

As shown in Figure 7.1, a *gateway* acts as an intermediary for a server. It's a gate through which all communications must pass. The most common example of a gateway is a corporate firewall that protects a corporate server from unauthorized access. It forwards communications from outside a corporation's network to servers on the network, but only if the programs initiating the communication have permission. The gateway's job, in this case, is to protect the server.

A gateway may use any number of protocols. In addition, the program trying to communicate with a server inside the firewall may not be aware that a gateway exists between itself and an Internet server.

Like a gateway, a *proxy server* also intermediates between some Internet-savvy piece of software, like a browser, and a server. Unlike a gateway, which operates for the benefit of a server, a proxy server's job is to make requests on behalf of other Internet client programs like browsers.

For instance, a company may not let an employee directly connect to the Internet. Instead, they might require that all requests to web sites outside the company be funneled through a corporate server. This lets the company control access to the outside. If an employee wants to access Yahoo!, their request is forwarded to the proxy server, which has access to both the internal company network and the Internet. It in turn makes the request to Yahoo!, and passes the results back to the employee.

Unless you are charged with actually writing a gateway or proxy server, you don't really need to understand the subtle differences between the two. You can use the two terms interchangeably. What you do need to know is that there's a computer between your WAP microbrowser and the content server. That computer provides critical functionality necessary to make WAP operate properly.

For more specific information on this terminology see the HTTP protocol description. [Fielding 1999]

WAP Gateway Services

WAP gateways provide a variety of services. It is up to software vendors to decide which features to include. Some of the most common features of a WAP gateway include:

Protocol Translation. Gateways are software language translators in much the same way interpreters translate between people who speak different languages. Many times, gateways speak multiple protocols to allow for flexibility. In Figure 7.1, the connection between the gateway and the Origin Server is an implied Internet connection. Therefore, the protocol being used is HTTP. WAP is designed to work with all popular wireless networks. Different gateway manufacturers support different networks.

Binary Data Encoding. HTTP is a text-based protocol. If you could tap into an Internet connection and watch the data go by, you could read the HTML data whizzing back forth between a web browser and a server. While text is

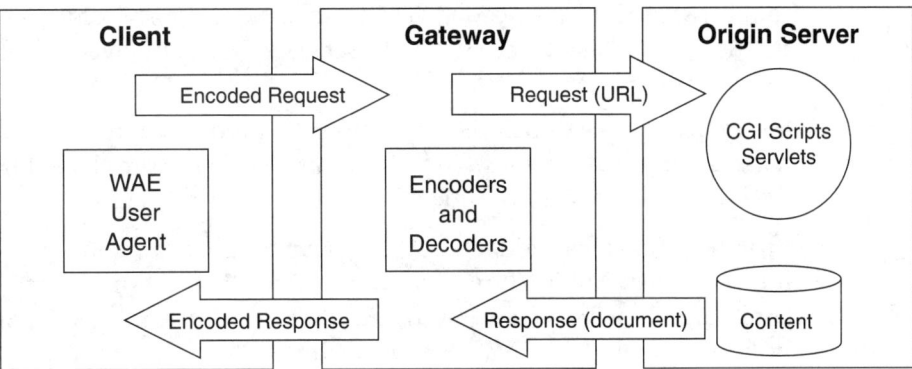

Figure 7.1 The WAP architecture.

an efficient way for humans to communicate, it's less than optimal and incredibly inefficient for computers. Text documents compressed to an equivalent binary format can occupy five to ten times less space. Wireless data transfer speeds currently range from 9.6 Kbps to 19.2 Kbps, much slower than dial-up wireline modems. In addition, wireless networks are notorious for interference and dropped connections, problems WAP is designed to deal with.

Unfortunately, the cost is that a greater portion of a 9.6 Kbps wireless connection is dedicated to error detection and correction. This makes the connection seem even slower than the equivalent wired connection. By taking data that is typically text-based and compressing it into a binary format, throughput can be much better. This encoding is a fundamental feature of all WAP gateways.

Translating HTML into WML. The great majority of WAP browsers do not directly support HTML. (Microsoft's Mobile Explorer is one exception to this rule. It handles both WML and HTML.) That makes sense since WML is designed for small-screen devices. When a gateway makes a request for a document from a web server on behalf of the WAP client, that web server should respond with a WML document. It's then up to the gateway to binary-encode the document and shoot it off to the WAP client.

However, there is no reason why a WAP gateway can't support the translation of other markup languages to WML before encoding and transferring a document. Phone.com's UP.Link gateway supports the binary encoding of HDML, HTML, and WML. They support HDML primarily for backward compatibility. There are a tremendous number of HDML-based services in existence. They support HTML because most web sites use it.

The majority of web pages out there are unsuitable or impossible to view on a WAP device. They contain Java applets, audio, video, JavaScript, screen layouts requiring a 1024 by 768 monitor, and so on. These media types and attributes are useless to WAP browsers. However, automatic translation from HTML to binary-encoded WML is a quick-and-dirty solution in some situations. Text-based web sites may translate reasonably well, but it's a trial-and-error process.

Gateway Software Extensions. No standard would be complete without manufacturer's extensions. These supplementary features let gateway creators differentiate their products.

For instance, Nokia has added an Application Programming Interface (API) to their gateway. An API is simply a set of functions that lets programmers do certain things. Nokia's gateway provides access to enterprise legacy systems without first going through an Internet or Intranet web server. This Java API lets enterprise IT developers program directly to backend systems and serve the data up as WML documents to a WAP device.

Other manufacturers offer application server suites to expand standard gateway functionality. They may provide services that gather e-mail from a variety of ISPs and corporate e-mail systems and present them to WAP clients. Other extensions handle calendar and contact information delivery from backend systems. Phone.com has a push feature that lets service providers send paging-like messages to WAP users.

Finding a Gateway

There are a variety of WAP gateway products on the market. We don't discuss them in detail here for two reasons. First, it's difficult to find accurate detailed information on the WAP gateways without initiating purchase discussions with the various vendors. Second, WAP gateway products are changing all the time. We didn't want to include any information in this book that might be obsolete before it was even printed.

We suggest three things. First, go to the WAP Forum web site and scour the membership information to identify the WAP gateway vendors. As of this writing, the WAP Forum's product and service guide is a bit outdated. Perhaps by the time you read this it will be more up-to-date.

Second, visit the various web sites we mention in the Resource Appendix at the end of the book. These third-party independent sites seem to do a better job than the WAP Forum of keeping up to date on current WAP products and services.

Finally, talk to the WAP Forum founders. Phone.com, Nokia, and Ericsson all offer gateway products, some of the most mature on the market. You probably can't go wrong with a product from any of these companies.

Once you have a list of potential vendors and products, match your requirements to their features and capabilities to see what might work best for you.

Security

For IT managers, security is a major concern. No technology lasts more than five minutes within a major corporation unless it has appropriate security capabilities. For wireless and mobile devices, security is an even more important issue. WAP devices are typically small and easy to steal. Since you can't adequately protect the physical security of such a device, issues such as user-authentication and privacy become paramount.

Privacy

To maintain privacy you must prevent people from eavesdropping on a conversation, reading e-mail that is not addressed to them, and intercepting other communications. In a computer security system, privacy is usually guaranteed by encrypting data. Data encryption requires sophisticated algorithms to scramble data such that it is unintelligible to people who should not view it.

Encryption algorithms come in a variety of strengths, dictated by the amount of computing power and time necessary to decode the transmission if you don't have the key. The strength of encryption algorithms is normally described in bits: 56-bit encryption, 128-bit encryption, and so on. The more bits, the stronger the encryption.

Integrity

Data integrity guarantees that the information you send is the same information that is received. If you send an electronic authorization wiring $10,000 dollars to someone, you would not be happy if an intermediary changed that amount to $1,000,000. A security system with guaranteed integrity would have a way to detect any changes to the original data.

Authentication

Another important security aspect is authentication. Authentication guarantees someone's identification. To authenticate yourself in real life you might show someone a driver's license or passport with a photograph. In computer land, authentication is done with digital certificates.

Digital certificates are issued by certificate authorities, groups, or companies to whom the requester has to identify him or herself. Once the digital certificate is issued and installed on a computer, that computer can use the certificate to authenticate itself to other computers. Client authentication guarantees the client identity. Server authentication guarantees the server identity.

For more information on digital certificates visit home.netscape.com/security/techbriefs/certificates/howcerts.html.

Non-Repudiation

Non-repudiation guarantees that a person truly decided to enter into a transaction. A photo identification authenticates a user to start a transaction; a signature on a document that can be matched to the photo ID ensures non-repudiation.

In computing, non-repudiation security comes in two forms: user name/password pairs and digital signatures. If no one else knows your password and the password cannot be easily guessed, using that password guarantees that you, and not someone else, is participating in a transaction. Digital signatures, on the other hand, have nothing to do with signing your name on a UPS driver's electronic tablet. They involve public key cryptography which is beyond the scope of this book.

You can find a tutorial on digital signatures on the American Bar Association's web site at www.abanet.org/scitech/ec/isc/dsg-tutorial.html. Another great source of information with examples is the Digital Signature Trust Company's web site at www.digsigtrust.com/pdsb.html.

WAP's Security

WAP communication is encrypted between the client device and the WAP gateway. Once the gateway receives and decrypts the data, another form of data encryption must be used to secure it over an Internet connection. The typical method for this doing this is through a Secure Sockets Layer connection, or SSL, between the gateway and the web server.

WTLS is used between the WAP client and WAP gateway. SSL is used between the WAP gateway and web server. Part of a gateway's job is to translate between the two security methods. By design, both SSL and WTLS handle authentication, privacy, and integrity. You need to institute another method like digital signatures or username/password pairs to enforce non-repudiation.

It can be overwhelming when trying to understand all the aspects of computing security. Millions of dollars have been spent figuring out solutions to these problems. There are numerous books written on the topic. Fortunately, WAP's creators thought about all these issues when designing the WAP protocols. They included solutions to the most critical security problems.

Just as web servers are critical to the success of the Internet, WAP gateways are in many ways the most crucial part of the WAP model. As WAP vendors get a feel for the types of services that users want and need, WAP gateway capabilities will grow substantially in the future. Today's gateways give you just the smallest glimpse of what the future holds.

CHAPTER 8

Some WAP Profiles

What value does WAP have if no one implements it? Fortunately, there are numerous consumer and enterprise deployments occurring around the world. The United States, where WAP originated, is lagging in cellular technology. Many countries in Europe and Asia have more widespread cellular infrastructures and WAP services than the United States.

In this chapter, we describe four WAP implementations: three enterprise and one consumer. The Holliday Group (www.holliday.co.nz), a New Zealand mobile and WAP software developer, was the key development partner for all three enterprise engagements.

exo-net

Company Name: exo-net (www.exo-net.com)
Location: Auckland, New Zealand
Project Name: Key Performance Indicators
Project Manager: Mark Loveys

Business Background

exo-net develops and distributes Enterprise Resource Planning (ERP) software solutions in Australia, New Zealand, and Singapore for small-to medium-sized enterprises. Their solutions are full-featured, 32-bit ERP applications with native interfaces to the World Wide Web. Many of their customers who have implemented the application suite have had little problem getting up and running on the web. WAP technology is very important to the future of this ERP suite and the next logical step for exo-net.

exo-net has gained considerable market share over the past 12 months and would like to continue to introduce leading-edge technology for its clients. Applications running on smart phones are viewed as important for the introduction of this core ERP product to the Singapore market. Singapore has been very aggressive rolling out WAP.

exo-net will be releasing a number of WAP applications over the next several months. The first application is access to Key Performance Indicators. It is intended for use by senior management as an executive information system.

Features

exo-net's first module, Key Performance Indicators, includes access to:

- Figures, sales, service incidents, and credits
- Report headers
- Bank details and current account status
- Creditor and debtor details

Previously this information was only available from a PC within the office environment.

This first application targets senior management, the top 10–15 percent of a client's management team. It is seen as a convenience tool for employees who already carry mobile devices and understand the power of quick access to critical information.

The Key Performance Indicators application is exo-net's first WAP application. It has intentionally been kept simple to shorten the development timeline and to ease the company into the mobile data arena.

WAP Background

There are strong GSM networks in exo-net's geographic territory. New Zealand has one GSM carrier. Australia and Singapore each have multiple carriers. All of these cellular companies operate WAP gateways to run WAP applications and have strong marketing programs to sell mobile data services.

exo-net considered extending their core product to accommodate handheld technologies. These programs will likely be developed, but at a later stage. This relatively quick and painless WAP solution is not part of their core product suite. However, it has a high return on investment and can be marketed to a wide range of customers.

Technology and Development

exo-net considered developing for two different WAP devices: the Nokia 7110 and the Ericsson R320 (see Chapter 6, WAP Client Software, Hardware, and Web Sites for more details on the 7110 and R320). These devices were recently released in this market. They are being widely accepted as wireless web access tools. In the end, exo-net decided to use both devices.

The Holliday Group, exo-net's software development partner for this project, offers software that runs on a Microsoft IIS web server and generates WML on the fly. The software creates dynamic WML content from the underlying data content. This is a benefit—the server can do more calculations and computations on the server side than it can when serving static content. Consequently, Holliday doesn't need to use any WMLScript for device-side calculations and can field a WAP program that uses just WML. The application can run on the lowest common denominator WAP device.

For development, the Holliday Group used Microsoft Windows NT Server with IIS with software telephone simulators. For testing they used an online web server with smart phones and WAP browser-enabled handheld devices. The application runs through both the Nokia and Phone.com gateways in New Zealand.

The development time was approximately two weeks. One engineer worked full time on the WAP application and implementation. The other engineer was responsible for the database integration. Both development and testing went as expected and did not delay the rollout of the service.

Project Status

As of May 2000, the application is just being deployed. The initial impression is that the data access is somewhat slow due to the circuit switched connection being used. Unfortunately, it can not be easily accessed in any other way. However, this application achieves both of the business goals. It serves as a learning experience in the WAP market and it works as a mobile, quick-access tool to critical management data.

MainFreight

Company Name: MainFreight (www.mainfreight.com)
Location: Auckland, New Zealand
Project Name: Freight Tracking
Project Manager: Garry Collings

Business Background

MainFreight is one of the largest long haul freight movers in Australia and New Zealand. They are one of the first freight companies to use the web for ordering and tracking freight business. To date, they have maintained a technology lead over their competitors.

MainFreight's key requirement is the ability to access tracking information on demand from any location. Their WAP application's purpose is two-fold: to raise the company's level of technology usage, and to give customers a higher level of service. The cost of tracking will only be lowered slightly since they expect a large number of customers to continue using the web interface.

Overall, the company expects increased revenue as new customers sign-on to use MainFreight due to their advanced technology. MainFreight has thousands of customers, but only about five to ten percent are expected to use this service in the first 12 months as the acceptance of WAP grows slowly.

WAP Background

There are significant GSM networks in the regions where MainFreight operates. New Zealand has one GSM carrier. Australia and Singapore both have multiple carriers. All of these cellular companies operate WAP gateways to run WAP applications and have strong marketing programs to sell mobile data services.

Technology and Development

MainFreight considered developing for two WAP devices: the Nokia 7110 and the Ericsson R320. Both of these devices had just been released in this market and were widely accepted as a wireless web access tool. They decided to target both devices.

MainFreight's WML application runs on their main Internet server along with the code to access the tracking database. The tracking database runs on a different server in the same physical location. MainFreight uses the cellular telephone company's gateway instead of hosting their own since this application requires little security. Only password access is necessary.

The application lets a customer enter a consignment/tracking number into the phone. It passes back to the customer the current location of the consignment. The application can show the end location, the receiving person, and a number of weigh points. It also provides a telephone number if the customer requires more information. This number gives customers immediate access to MainFreight's national call center.

The development time on this project was two weeks. MainFreight supplied both the engineering and project management skills to the project, and used one Holliday engineer.

Project Status

Like exo-net, MainFreight is in the earliest stages of rollout, so it's difficult to tell how the service will be received. The company is testing the WAP waters with this relatively easily-to-implement application. They're optimistic about the outcome.

Sky City Hotels

Company Name: Sky City Hotels (www.skytower.co.nz)
Location: Auckland, New Zealand
Project Name: Hotel Room Booking and Checkout
Project Manager: Tim Partington

Business Background

Customers of Sky City Hotels, an upscale hotel/casino chain, use this application. The goal is to let customers of the hotel chain make reservations and modify them at will. It is also supposed to speed the checkout process. Sky City Hotels is targeting the service to its regular, high-spending customers. The hotel chain regularly uses technology to maintain its reputation as a leading service provider in the industry. Sky City Hotels views this as added value for the customer as well as a technology showcase to woo potential customers.

WAP Background

The D-AMPS (CDPD) network in New Zealand was chosen for this application. The Ericsson R280 is the only telephone available for this network. It supports voice and a 19.2-Kbps data rate over a packet network, which is much more suitable for WAP than a circuit switched connection. The coverage footprint for the D-AMPS service in New Zealand is the only issue with the network.

Sky City decided to use WAP for this trial to determine if the technology is suitable for this type of application. The program lets an existing client request a hotel room reservation from his mobile phone. The room confirmation is passed directly back to the device. If required, the client can press a button on the phone to directly call Customer Service. Additionally, a client may review a current booking, change a current booking, and make a request to checkout.

Technology and Development

The Ericsson R280 supports HDML, the progenitor to WML. Phone.com supplied the gateway technology for the application.

Security is not an issue for this application. The user is associated with a mobile phone number that acts as a security key. HDML is served on the fly from the application server. The cellular telephone company provides the gateway for two reasons. First, Sky City Hotels views this implementation as a trial service. Second, the hotel does not see added benefit in hosting their own gateway.

Development time on this project was about one month with two engineers working on the project. Most of the implementation time was spent streamlining the information coming from the database. Skills important to this mobile implementation, as well as the two projects previously mentioned, include WML (and HDML), WMLScript, generic programming skills, and an understanding of TCP/IP networks to enable infrastructure setup.

Retrospect

The application is currently being deployed.

A Consumer Profile

Person: Scott Sbihli

Location: Midwest America

Project: Consumer HDML/WAP Services

In the United States in May of the year 2000, there are few WAP services. Most wireless services are deployed using HDML, WAP's predecessor technology. Two major rollouts of WAP technology include services from Sprint PCS and Verizon Wireless.

In April, Verizon turned these services on for consumers in Ohio. The smart phone we used to connect to these CDMA-based services is the QUALCOMM QCP-860 with an integrated Phone.com UP.Browser.

Services

Obtaining access to Verizon's Mobile Web requires two button clicks. Since the phone has a unique ID and a user is associated with that ID, you gain immediate access to the Internet. The phone authenticates you over an encrypted connection using RSA security, which prevents eavesdropping. Most cellular packages today come with airtime minutes included in the monthly service fee. Using data services simply counts against these minutes.

Once connected, a menu containing ten options appears:

My.Airtouch. My.Airtouch extends to WAP devices, services Verizon Wireless provides its customers on their web site. They include calendaring, e-mail, an address book, news, sports, stocks, and weather information. The calendar can be shared among several people. Most of the functionality available on the web site can be accessed using the UP.Browser on the smart phone. While not very practical due to limited input abilities and screen size, you can add, edit, and delete address book and calendar entries from the telephone in addition to viewing the information.

Messaging. You can send both Airmail and e-mail from the telephone. Airmail lets you message alphanumeric pagers and cellular phones. To send Airmail, you just type in a telephone number and then compose your message with the keypad. Without some type of advanced text entry feature, the task becomes tedious. E-mail works much the same way, but you have the option of pulling e-mail addresses directly from your personal address book stored at Verizon.

Hotlist. This feature includes Microsoft's MSN Mobile Services, and a Hotlist that contains ski reports and movie listings. The movie list is searchable in a variety of ways including by city and movie title. The service has links to your personal Airtouch profile so that when you search, your home city is used as the default. Unfortunately, it takes many button clicks to drill down to a point where you can actually retrieve movie times and descriptions. A customization service for this feature would be of great help, such that a one-button click would tell you the movie times that night for your favorite local movie theater.

Directories. Verizon offers White Pages, Yellow Pages, and a Reverse Lookup for finding information based on a telephone number. Once you locate a person or business, a few clicks dials their number or gets driving directions to their location. By default the driving directions start from the address set up in your web site. You can add any information retrieved to your address book and use it later from the telephone or web site. Without a doubt this is one of the service's most useful features.

Financial. These services are not yet implemented.

Shopping. Currently you are limited to shopping at Amazon.com. You can purchase books, music, videos, DVD disks, toys, games, electronics, and home improvement products. There are a number of methods for finding products. Once you register yourself with the service the product-buying process becomes easier and quicker.

On The Road. On The Road lets you find hotels, movie theaters, gas stations, and auto repair centers, and obtain driving directions.

Bookmarks. If you bookmark different web sites, this is where you find them.

GoTo... This feature lets you enter a URL, any URL. If you surf to a site, such as Yahoo!, with HDML or WAP services, you get those services. Type in a more generic URL and the Verizon data service tries its best to present the site to your WAP phone in some format that's readable.

Customer Care. From this option you can view Frequently Asked Questions about the data services, read help files, and call Customer Care to ask for assistance.

Theory versus Practice

Without a doubt, wireless data services are addictive. The ability to check e-mail, review movie times, and read the weather forecast can be useful wherever you are. As WAP gains popularity, providers will learn which services are most useful to subscribers. For example, do consumers really have a need or a desire to purchase books or movies while standing in line at a McDonalds?

Additionally, WAP application developers and device designers will learn how to build better user interfaces. For example, if you frequently request driving directions, service providers should place that item on the first card of your home deck when you connect. In addition, it should take only a few key presses to extract the information. As WAP devices get larger screens, better resolution, color, and more processing power, you may find your regular text driving directions delivered with a color map and an interface to a Global Positioning System.

When compact discs first became popular as a media for distributing software, many manufacturers simply repackaged the floppy disk versions of their software. This process became known as Shovelware because companies hastily shoveled content onto the compact disc without much thought as to the advantages and disadvantages of the delivery media.

The Verizon WAP services have a similar feeling. There's a lot of content, but some of it is simply too difficult to access quickly without entering alphanumeric characters. This presents a challenge without some type of advanced data input like voice recognition or a touch screen.

Just like web portals like Yahoo! let you customize your experience with their site, you can expect similar capabilities from WAP providers. If you never read the news, why should you have to navigate past that item each time you use your telephone? If reading a horoscope is a top priority for you in the morning, you should be able to assign it to a hot key for easy access.

What WAP Does Well

In all four of the profiles in this chapter, WAP is used to retrieve and send very specific types of Internet-based content. Custom application developers created solutions to access ordering and booking information, and executive level reports. Consumer service providers take a similar approach. They filter out the majority of Internet content and present what they feel is most useful to mobile consumers.

While Verizon's implementation lets you type in any URL to retrieve content, it's not a well-advertised feature because of the substantial limitations of current services to present content that is not specifically designed for WAP devices. As with any technology, developers and service providers will become more adept at pinpointing the best information and the correct methods for its presentation.

CHAPTER 9

Implementing an Enterprise WAP Strategy

A typical enterprise IT staff will have a person or group that evaluates new technologies for the company. They are the gatekeepers that decide which technologies deserve further investigation and implementation because of their business value. They also decide which technologies are simply hype, fads, or a series of undelivered promises.

Once a technology makes it through the door, the IT staff must make all of the promises in the marketing materials a reality. Regardless of the technology and the experience of the IT staff, each new implementation poses its own set of challenges, nuances, and quirks.

WAP is no different.

In this chapter we provide advice to IT managers to help them avoid WAP implementation potholes. There are numerous software project management methodologies. We don't recommend any particular one. Instead, we discuss the aspects of the software project lifecycle that WAP is likely to affect so that you can make the necessary adjustments to have a successful WAP implementation.

Requirements

Gathering functional requirements for a WAP application is different than for a typical desktop or web application. WAP devices simply have fewer resources than desktop computers—smaller screens, less memory, slower networks, and less processing speed. While mobile devices will become more powerful in the coming years, they will never catch their desktop counterparts. Therefore, it becomes crucial to design for these limitations.

It's easier to describe what doesn't work on a WAP device than it is to make a list of things that do. Any application that requires substantial horizontal or vertical scrolling does not function well. Applications that need to display several columns of data fall short. Software designers must work hard to give the end user worthwhile information in a mobile sense. The information you need while working on a desktop computer is simply different from the data needed on a WAP-based device.

For example, let's say that the goal for your WAP application is to gather flight arrival and departure information from the major airlines' web sites and reformat the data into WML for your company's employees to use. The desktop version of the application could display the scheduled arrival and departure times, the actual arrival and departure times, the airlines' logos, flight numbers, food choices, and so on. While many people would state that all of this data is crucial, including the name of the in-flight movie, the reality is that the WAP version would strip out the majority of this information. The most relevant information for business travelers is arguably the actual departure and arrival times.

What it comes down to is managing user expectations. Most users' expectations have been driven by the web for so long that they only think in terms of the web paradigm. This typically includes a long, scrollable page with media rich content. Instead, WAP deals with multiple cards and navigation between these cards. It's important to build an expectation based on the WAP deck/card paradigm in which much smaller pieces of information appear at a time. You must concentrate on delivering the information without any extraneous niceties.

Architecture

When designing WAP solutions it's critical to separate the content shown on a page from how that page is presented and formatted. We look to HTML as a good example of how not to structure information. HTML lets developers completely intermix content with the content's presentation. Typically, in an HTML file you'll find a piece of data for display, and immediately to either side of that data are tags that describe any number of attributes that data should acquire: Bold, Arial, hotlinks, and so on. Here's an example:

```
<html>
 <head>
  <title>WAP</title>
 </head>
 <body text="Red">
  <h1>WAP Overview</h1>
```

```
    <p>This page is <I><B>the</B></I> definitive guide to WAP technology.
  </body>
</html>
```

The word **the** in the sentence has `<I>`, `<B>`, `</B>`, and `</I>` tags that dictate how it should be displayed. In this case, the word **the** is shown italicized and in bold. If you need the information in a different format—say in WML—it can be difficult to extract the data and reformat it for a new markup language.

This scenario is becoming more of a reality everyday. More and more businesses need to deliver their content through several different channels. Examples include the web, WAP, interactive television, and e-mail, to name a few. As consumers and businesses request the data in these different formats, the data providers naturally wish to minimize their development effort.

Fortunately, there is a general movement toward solving this problem by separating content and presentation. XML is one way of storing content. In the future, this XML data can be presented in any number of ways using different presentation methods. WAP then becomes the presentation layer for small, resource-limited devices.

WAP enables the delivery of data to mobile users. However, that data is merely a sub-set of a larger data set that an enterprise maintains. The company may choose to deliver the same data plus additional items not suitable for mobile devices, such as graphics, to a user sitting at a desktop.

Don't view a desktop delivery solution and a mobile delivery solution as two completely separate development channels. Rather, see them as three smaller ones. The database, its queries, and reports are one development path. The other two are the presentation of this data to two different user profiles: a desktop user and a mobile user. Figure 9.1 shows this graphically. Once the data is properly designed, it's much simpler to extract the correct data for presentation to any type of device. Spend time early on configuring the data to accommodate a variety of current and future presentation methods.

Taking the time to go through this exercise is an advantage because it ultimately reduces total development time. When you need to support a new presentation method, you can do so with limited effort, rather than forming a new development team to convert both the presentation method and the data.

Design/Implementation

Once you gather your requirements and select an architecture, design, development, and testing start. Many of the skills leveraged for these latter stages are similar to traditional web development. However, here are some tips for creating better WAP solutions:

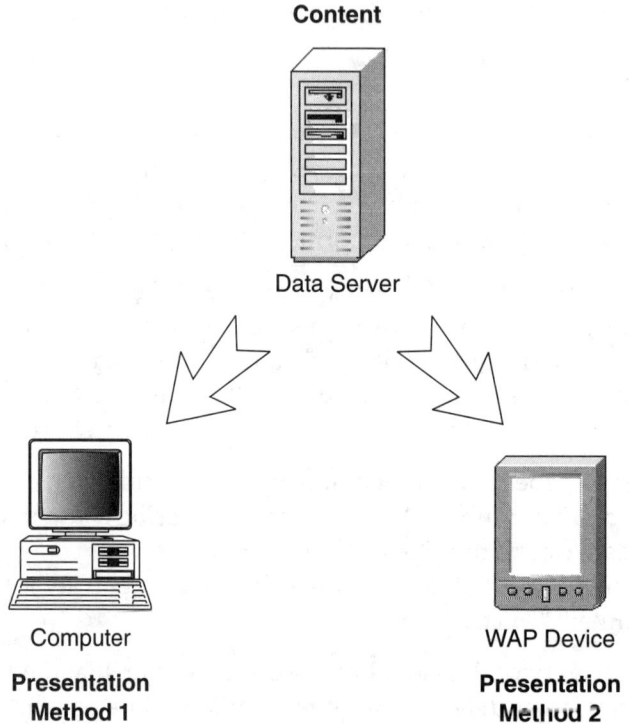

Figure 9.1 Separate presentation from content.

- ➤ Determine the user interface and navigation model as early as possible. Will you use option screens, programmed hardware buttons, or does the application require a voice recognition interface? User interfaces are limited on WAP devices. Also, there is no guarantee that an input method on one device will be available on another.

- ➤ Keep the depth of cards to a minimum. Navigation within a traditional web browser is easy. A user can see the URL they're viewing, they have a mouse to quickly click links and navigate pages, and web designers can graphically or textually show a user where they are within a site. WAP users do not have these interface and contextual advantages. A poorly designed application results in lost, frustrated users who repeatedly press the End button to exit from the microbrowser.

- ➤ Include WAP-specific features. If the telephone supports voice recognition, a touch screen, or integrated telephony features, determine if and how they can benefit the user and his experience. Make sure that the users all use the same device or you lose these advantages.

- ➤ Specific WAP presentation features differ among devices. As we saw in an earlier chapter, some WAP devices support bold text while others do not.

Don't rely on these optional features to render your data if the style of presentation is critical to the data interpretation.

➤ Data caching is usually fundamental to the speed of your application, provided you aren't serving up real time stock market quotes. If the data is not going to change often, then design your application so that your decks are cached for a long time.

➤ Look at the size of an application's decks and cards. Keep decks to a size of one kilobyte or less. Keep in mind that there is most likely more than one way to accomplish what you trying to do.

➤ Different wireless networks perform differently. Circuit switched networks require a longer connect time than packet switched networks. A circuit switched network is one where a device has to dial-in before it can transfer data. A packet switched network is similar to a paging service. No connect time is necessary.

➤ Keep in mind that wireless networks typically have data rates between 9600 bits per second (bps) and 19,200 bps. This raw data rate is normally much greater than the actual data throughput. Wireless protocols like WAP have significant overhead built-in to the protocol to handle error detection and correction.

Outside of these tips, Phil Holliday of the Holliday Group (www.holliday.co.nz), a New Zealand mobile and WAP software developer, offers the following advice to WAP implementers:

"All the WAP development we have done so far has been based on backend data systems that were in place. We've also used gateways provided by different cellular companies. With that in mind, WAP development has not been difficult. Rather, the challenge has been managing client expectations because these are the initial days of WAP. We are dealing with the early adopters who wish to make a statement about their business by embracing new technology. This is not to say that the applications are trivial or useless, but we expect to see the real business advantage being made in the WAP market over the next two years.

"If you are an IT professional, consider what technical skills you will need that you do not currently possess. In many ways, you'll find that in managing WAP projects, nothing much changes. The rules stay the same: Define the requirements, design the system, build the application, run it on a test system—NEVER on the live system—and deploy the application.

"If you need to install your own gateway, you have a new technology to learn. Will you go to an industry leader, such as Phone.com, a telephone manufacturer, a third-party developer, or a freeware gateway solution? (Yes, there are public domain gateways out there right now.)

"You will need some additional tools. In particular, you must obtain as many phones or phone simulators as you can get to ensure that your application looks as good as possible. You will need a WAP development toolset. This could be

Cold Fusion or similar, or one of the free development environments from the phone manufacturers."

Testing

Test your solution on as many different devices as possible. Each device has its own particular method of displaying WML pages. Some telephones support graphics, italicized text, and on-screen navigation. Others support bold text, touch screens, and hardware button navigation. If the application will be used by one business or a small group across business units, try to standardize on a single WAP device to reduce your support and training costs.

Test your application against multiple bearer networks and gateways, if possible. Gateway feature sets vary greatly across vendors. WAP technology is still relatively new and the same application may perform very differently between two gateways. Make sure to evaluate the gateway against your application before committing to one. Application performance may also perform differently depending on the type of network it's being run on.

Summary

WAP is still is in its infancy. Businesses in Europe and Asia have started deployments because of the prevalence of WAP-based devices. The United States lags behind, but will catch up rapidly during the next few years. Managing WAP development projects is not that much different from typical software project management, but it's not the same. Best practices are evolving daily. Expect to see certain design and programming practices become more prevalent than others.

Keep the following key points in mind:

- *Get your backend data stores in order.* Find a storage method such as XML that separates the content from the presentation. If you do, you'll be prepared to deliver content to any current or future presentation device.
- *Manage user expectations accordingly.* Mobile devices have different capabilities than their desktop or laptop counterparts. Deliver only necessary data.
- *Keep the application and its navigation simple and intuitive.* It is easy for a user to become lost within an application and give up on it.
- *Test, test, test.* Test across gateways. Test across devices. WAP's flexibility in microbrowser implementation demands this. It's the only way to guarantee consistent user experiences.

CHAPTER 10

The Future of WAP

WAP is still in its infancy, and the jury is out as to whether it will be successful. Some industry experts predict that within five years, all cellular telephones will include WAP browsers and there will be thousands of WAP-compatible information services. Other experts don't think WAP provides any compelling advantages over existing Internet technologies.

In this chapter we discuss problems with the current WAP implementations and future hardware and software directions that may have an impact on WAP's popularity.

Problems with WAP

The first time you use a WAP-enabled telephone, its problems from a user perspective are immediately obvious.

Screen Size

The first is screen size. You just can't fit very much information on a screen that is four lines by 15 characters. For anyone who regularly uses a web browser on a desktop computer, the small screens are particularly annoying. Cellular telephone manufacturers are doing a good job making crisp, clear, readable screens, and including icons and graphics to depict more information in less space. However, that doesn't solve the problem. It just reduces the pain.

Navigation

The second problem is navigation. Due to the small screen size, you have to spend more time moving from screen to screen and line to line within a screen than you do on a desktop system. Suppose an application wants to present you with a list of nine choices so that you can pick one. On a desktop system, those nine items might appear as nine check boxes. All you have to do is click your mouse on one of them. Alternately, they might appear as a pop-up or drop-down list. You have to click on the list once, then click again on your choice.

On a small-screen WAP telephone, you can't even see the nine choices, only the first three or four. You have to press some sort of scroll button to move down the list to see the rest of the choices. Then, you have to press the scroll button a few more times to position the cursor on your choice, then select it. The WAP telephone may take five to 10 actions to do the same thing that takes one or two actions on a desktop.

Data Entry

The third problem is data entry. Cellular telephones don't have keyboards, mice, or voice recognition for data entry. The most common way to enter alphanumeric text is by repeatedly pressing one of the telephone keys. For instance, if you want to enter the letter "C," you have to press the "2" key four times to cycle it through the characters "2," "A," "B," and "C." You then pause for a second so that the microbrowser can move the cursor to the next character position before entering the second character.

This restriction makes it impossible to enter quickly and easily text on a WAP telephone.

Latency

Latency is a measure of the amount of time it takes for something to finish. For instance, when you request a web page using a desktop browser, the latency is the amount of time it takes for the full page to appear on the screen. Your desktop latency is affected by a variety of factors, including the speed of your CPU, the amount of RAM you have, the speed of your Internet connection, the load on the server delivering the information to you, and more.

WAP devices also experience latency. Because wireless communication links are typically much slower than wired connections, it can take several seconds for a screen of WAP information to appear. In addition, since every WAP page is much smaller than a web page, chances are that in order to do anything useful on a WAP telephone, you need to fetch several pages.

The net result is that it usually takes more time to do something with a WAP telephone than with a desktop system or even just a telephone. There is an often-repeated anecdote that if you want to check the arrival time of a flight using a WAP telephone, you should call the airline. A variation on that anecdote is that by the time you can enter an ISBN number on a WAP phone to order a book from Amazon, you could find the closest Barnes & Nobles and buy it in person.

> **NOTE** Many of WAP's shortcomings as described in the preceding sections can be circumvented, to a certain extent, by clever application development. Well-written WAP programs minimize navigation, data entry, and latency. Even the best programmers cannot compensate for small screen size, however.

Duplicate Content

WAP also has a serious problem from a content developer's perspective. Most content providers already have a web presence. Many want to migrate their content to WAP. As we've described, web sites are typically built using HTML and HTML-related languages. WAP sites are built using WML and WMLScript. Content developers that want to support both environments have to create two separate sets of content.

WML is similar to HTML, so it's not too difficult to support both types of content. However, it does take extra work. In the fast-paced world of the Internet, any extra work is a burden.

Solving These Problems

We don't intend to paint a bleak picture of WAP. In fact, it's likely that most of these problems will be eliminated over the next few years. Here's how.

Screen Size

Cellular telephones have small screens for two reasons: tradition and cost. In the past, standard telephones have not really needed big screens because there wasn't much to look at other than, perhaps, the number you were dialing and a signal strength indicator. People have grown accustomed to small screens and seem to prefer the small telephones they fit in.

Small screens are also less expensive than large screens and require less power, keeping battery life reasonable. Most people prefer to pay as little as possible for a telephone, and they want the longest battery life.

There are no silver bullets on the horizon that will eliminate the screen size problem. Like all facets of computer technology, however, manufacturers are constantly improving their wares. Screen costs are going down, new screen technologies are being developed, and battery life is improving. You can now buy a $200 Palm device that has a 160 × 160 pixel touch-sensitive screen and battery life that extends to weeks. We should see cellular telephones with larger screens within the next twelve months.

Navigation

Once screens get larger, navigation improves. More information can be put on each screen and touch-sensitive capabilities make it easier to make choices and select elements on a screen. In addition, manufacturers are working on improved navigation mechanisms. For instance, just within the past twelve months a gadget called a thumb wheel has appeared on some devices. It lets you move through a group of choices just by rotating a horizontally-mounted wheel (see Figure 10.1). Expect more innovative hardware widgets in the future.

Data Entry

Data entry is a tougher problem as long as telephones have only keypads. Some improvements have appeared. One is predictive data entry. As you enter a character, the device tries to predict what you might want to enter next. It displays a list of choices on the screen so that you only need to select one of those choices instead of completely entering the rest of the word.

As we've already described, larger touch-sensitive screens should eliminate the navigation problems. Coupled with keystroke recognition software similar to Palm Inc.'s Graffiti, they should also make data entry much easier.

Figure 10.1 A thumb wheel navigation control.

Latency

The final problem is latency. Once again, hardware is part of the solution. First, processors are getting faster without impacting battery life. Faster processors mean faster program execution, so that delays are shorter.

Second, memory is constantly getting smaller, faster, and less expensive. Current cellular telephones typically contain on the order of a few hundred kilobytes (thousands of bytes) of memory for storing programs. Within the next 12–18 months, that will increase by at least a factor of ten. More memory means more space for programs and data that no longer have to be fetched from a wirelessly connected web site.

Third, we've already briefly discussed the next generation wireless telephone systems that are starting to be tested in various parts of the world. Wireless data speeds will get much faster in the next few years, making wireless Internet access as fast as wired Internet access for individuals. When that happens, the latency problem disappears.

The Next Generation

WAP was originally designed to deliver Internet-hosted content to small, low-power, wireless devices with limited memory, processing power, and screen size. As we discuss in the previous section, however, most of the limitations that WAP was designed to circumvent will be going away.

We strongly believe that within the next few years, cellular telephones will become much more capable. They will have bigger screens, better data-entry mechanisms, faster connections, more memory, and much more sophisticated software. They will still be small and light enough to fit in your pocket without ripping its seam. Essentially, they will be general-purpose programmable computers that also happen to have seamlessly integrated wireless voice and data capabilities.

Once that happens, WAP's value starts to come into question. It solves wireless data problems that won't exist in a few years, and does so with questionable utility. Content providers and application developers currently have to do extra work to build WAP-based solutions. When devices become more robust, robust enough to support real web browsers, they may not want to continue to write two versions of their applications.

APPENDIX A

Resources

Documents

This brief bibliography includes books, magazine articles, and electronic documents from web sites. For electronic documents, we provide a URL where you should be able to find the document. Please note that web site names do change over time. We apologize in advance if any of the URLs are incorrect.

Bray, Tim, Jean Paoli, and C. M. Sperberg-McQueen, eds. "Extensible Markup Language (XML) 1.0", W3C Recommendation 10-February-1998, REC-xml-19980210. www.w3.org/TR/1998/REC-xml-19980210.

Brodsky, Ira. 1977. *Wireless Computing: A Manager's Guide to Wireless Networking*. John Wiley & Sons, Inc., ISBN 0-471-28656-7.

Fielding, R., et al. June 1999. "Hypertext Transfer Protocol - HTTP/1.1." www.w3.org/Protocols/HTTP/1.1/rfc2616.pdf.

Flanagan, David. 1998. *JavaScript, the Definitive Guide*. O'Reilly & Associates, Inc., ISBN 1-56592-234-4. www.ora.com.

"HDML Language Reference: Version 4.0." Unwired Planet. January 1999. www.phone.com.

Kunii, Irene M., and Stephen Baker. January 17, 2000. "Japan's Mobile Marvel." *Business Week*, p. 88.

Lamb, George. 1977. *GSM Made Simple*. Cordero Consulting. ISBN 0-966-57520-2.

APPENDIX A

Mann, Steve. 1999. *Programming Applications with the Wireless Application Protocol*. John Wiley & Sons, Inc. ISBN 0-471-32754-9.

Naik, Dilip. 1998. *Internet Standards and Protocols*. Microsoft Press. ISBN 1-57231-692-6.

Official Wireless Application Protocol: The Complete Standard with Searchable CD-ROM, Wireless Application Protocol Forum, Ltd., John Wiley & Sons, Inc. 1999. ISBN 0-471-32755-7. www.wiley.com/compbooks/WAP.

Raggett, David, Arnaud Le Hors, and Ian Jacobs, eds., et al. 1999. "HTML 4.01 Specification, W3C Recommendation 24-December-1999." www.w3.org/TR/REC-html40.

Standard ECMA-262: "ECMAScript Language Specification." ECMA, June 1997. www.ecma.ch/ecma1/PUBLICAT.HTM.

Stetz, Penelope. 1999. *The Cell Phone Handbook*. Aegis Publishing Group. ISBN 1-890154-12-1.

St. Laurent, Simon. 1999. *XML: A Primer*, Second Edition. IDG Books Worldwide. ISBN 0-764-53310-X.

"Understanding Security on the Wireless Internet: How WAP Security is Enabling Wireless E-Commerce Applications for Today and Tomorrow." Phone.com, 2000. www.phone.com/pub/Security_WP.pdf.

Ungar, David, and Randall B. Smith. October, 1987. "Self: The Power of Simplicity." OOPSLA '87 Conference Proceedings, 227-241, Orlando, FL.

"vCalendar—The Electronic Calendaring and Scheduling Format, Version 1.0." The Internet Mail Consortium (IMC), September 18, 1996. www.imc.org/pdi/vcal-10.doc.

"vCard—The Electronic Business Card; Version 2.1." The Internet Mail Consortium (IMC), September 18, 1996. www.imc.org/pdi/vcard-21.doc.

Web Sites

There are hundreds of WAP-related web sites. We don't try to identify them all. Instead, we include key sites that you can use as a starting point for investigating WAP. Many of these sites include links to other sites. Because of the transient nature of the Internet, some of the sites in this list may not exist by the time you read this book. They may have been renamed, taken offline, or absorbed by some other site. Regardless, there should be plenty of pointers here to get you started.

You should also check the last section, Consumer WAP Sites, in Chapter 6, WAP Client Software, Hardware, and Web Sites, for an annotated list of consumer-oriented WAP sites.

African Cellular

WAP is a global standard, but Africa does not get the same visibility as other continents. African Cellular, a retail company, offers all types of information and news about mobile technology, including WAP, with a special focus on Africa. www.cellular.co.za.

AllNetDevices

This is a premier site for news and information on handheld, mobile, and wireless devices. www.allnetdevices.com.

Anywhereyougo.com

Anywhereyougo is an excellent wireless development Web site with a plethora of content. www.anywhereyougo.com.

dmoz – Open Directory Project

dmoz is a search engine categorized by humans and not a web robot. They've created links to hundreds of Internet WAP resources. www.dmoz.org/Computers/Mobile_Computing/Wireless_Data/WAP/.

Ericsson

Ericsson played a founding role in the creation of the WAP Forum. They manufacture and design WAP-based telephones, handheld devices, software development kits, and gateway server software. www.ericsson.com/wap.

Gaddo.net

Gaddo contains articles, source code, tutorials, and resources dedicated to WAP development. www.gaddo.net.

Gelon.net

Gelon is a well-maintained WAP search engine. If you want to find links to WAP-based sites for your WAP device, this is the place to go. www.gelon.net.

GSMData

The GSM data web site is a clearing house for all organizations worldwide that are involved in implementing GSM-based wireless services. It contains technical, historical, and market information on GSM deployments. www.gsmdata.com.

mobileID

Like MyYahoo! does for regular web pages, mobileID lets you create your own custom homepage that can be served up to your favorite WAP device or simulator. www.mobileid.com.

Nokia

Nokia is another WAP Forum founder. They make cellular telephones as well as software infrastructure components (software development kits, WAP gateways, and so on) to support WAP. www.nokia.com/wap.

PCS Data

This web site is a clearing house for all things related to North American PCS communications. www.pcsdata.com.

Phone.com

Phone.com is one of the original founders of the WAP Forum. Much of the technology being used in the WAP specification comes from Phone.com. www.phone.com.

QUALCOMM

As the inventor of CDMA, QUALCOMM's web site has a great deal of technical and marketing information about CDMA, both the current generation and the future incarnations. www.qualcomm.com.

WAP Forum

Go back to where it all started. Browse the list of members and download the latest WAP specification from here. www.wapforum.org.

WAP Hole Sun

WAP Hole Sun is another WAP information repository. It has lots of links, news, and discussion groups dedicated to WAP. www.wapholesun.com.

WAP-Resources.net

This is an up-to-date site for WAP developers, although much of the content is of interest to non-developers. You can find a list of WAP companies, news, information on security, gateways, software development kits, and reviews. www.wap-resources.net.

The Wireless Data Forum

The Wireless Data Forum is an industry non-profit organization dedicated to promoting the use of wireless data. Their web site contains resources related to WAP and other types of wireless data. www.wirelessdata.org.

Wireless Developer Network

This site contains all types of information, news, and resources dedicated to developing wireless applications. You can also find an extensive library of links to other sites. www.wirelessdevnet.com.

Wireless Week

Wireless Week is an online newspaper covering all aspects of business and technology related to wireless communication. www.wirelessweek.com.

wow-com

wow-com is a web site run by the CTIA (Cellular Telecommunications Industry Association). It contains a wealth of information about various aspects of wireless communications, plus a daily news service that delivers wireless news via e-mail. www.wow-com.com.

Yes2WAP

Yes2WAP contains a variety of WAP overview information. Of particular interest are the links to WAP gateway makers, content suppliers, and network portals. www.yes2wap.com.

YourWAP.com

YourWAP is another customized portal site. Store contacts, e-mail, shopping list, and so on for viewing while on the road from your favorite WAP device. www.yourwap.com.

APPENDIX B

WAP WML

Proposed Version 19 February 2000

Wireless Markup Language Specification

Contents

1. Scope   **115**

2. Document Status   **115**
    2.1 Copyright Notice   115
    2.2 Errata   116
    2.3 Comments   116

3. References   **116**
    3.1 Normative References   116
    3.2 Informative References   117

4. Definitions and Abbreviations   **117**
    4.1 Definitions   117
    4.2 Abbreviations   118
    4.3 Device Types   119

5. WML and URLs   **119**
    5.1 URL Schemes   120
    5.2 Fragment Anchors   120
    5.3 Relative URLs   120

6. WML Character Set   **120**
    6.1 Reference Processing Model   121
    6.2 Character Entities   121

7. WML Syntax   **122**
    7.1 Entities   122
    7.2 Elements   122
    7.3 Attributes   122
    7.4 Comments   123
    7.5 Variables   123
    7.6 Case Sensitivity   123
    7.7 CDATA Section   123
    7.8 Processing Instructions   124
    7.9 Errors   124

8. Core WML Data Types   **124**
    8.1 Character Data   124
    8.2 Length   124
    8.3 Vdata   124
    8.4 Flow   125
    8.5 HREF   125
    8.6 Boolean   125

8.7 Number 125
8.8 xml:lang 125
8.9 The id and class Attributes 125
8.10 ContentType 126

9. Events and Navigation 126
9.1 Navigation and Event Handling 126
9.2 History 126
9.3 The Postfield Element 127
9.4 The Setvar Element 127
9.5 Tasks 128
    9.5.1 The Go Element *128*
    9.5.2 The Prev Element *131*
    9.5.3 The Refresh Element *131*
    9.5.4 The Noop Element *131*
9.6 Card/Deck Task Shadowing 132
9.7 The Do Element 133
9.8 The Anchor Element 135
9.9 The A Element 136
9.10 Intrinsic Events 137
    9.10.1 The Onevent Element *139*
    9.10.2 Card/Deck Intrinsic Events *139*

10. The State Model 139
10.1 The Browser Context 139
10.2 The Newcontext Attribute 140
10.3 Variables 140
    10.3.1 Variable Substitution *140*
    10.3.2 Parsing the Variable Substitution Syntax *142*
    10.3.3 The Dollar-sign Character *142*
    10.3.4 Setting Variables *142*
    10.3.5 Validation *143*
10.4 Context Restrictions 143

11. The Structure of WML Decks 143
11.1 Document Prologue 144
11.2 The WML Element 144
    11.2.1 A WML Example *144*
11.3 The Head Element 145
    11.3.1 The Access Element *145*
    11.3.2 The Meta Element *146*
11.4 The Template Element 147
11.5 The Card Element 148
    11.5.1 Card Intrinsic Events *148*
    11.5.2 The Card Element *148*
11.6 Control Elements 150
    11.6.1 The Tabindex Attribute *150*
    11.6.2 Select Lists *151*
    11.6.3 The Input Element *156*
    11.6.4 The Fieldset Element *159*

11.7 The Timer Element — 160
   11.7.1 Timer Example — *161*
11.8 Text — 162
   11.8.1 White Space — *162*
   11.8.2 Emphasis — *162*
   11.8.3 Paragraphs — *164*
   11.8.4 The Br Element — *165*
   11.8.5 The Table Element — *165*
   11.8.6 The Tr Element — *167*
   11.8.7 The Td Element — *167*
   11.8.8 Table Example — *167*
   11.8.9 The Pre Element — *168*
11.9 Images — 168

12. User Agent Semantics — 170
12.1 Deck Access Control — 170
12.2 Low-Memory Behaviour — 170
   12.2.1 Limited History — *170*
   12.2.2 Limited Browser Context Size — *170*
12.3 Error Handling — 171
12.4 Unknown DTD — 171
12.5 Reference Processing Behaviour - Inter-card Navigation — 171
   12.5.1 The Go Task — *171*
   12.5.2 The Prev Task — *172*
   12.5.3 The Noop Task — *172*
   12.5.4 The Refresh Task — *172*
   12.5.5 Task Execution Failure — *173*

13. WML Reference Information — 173
13.1 Document Identifiers — 173
   13.1.1 SGML Public Identifier — *173*
   13.1.2 WML Media Type — *173*
13.2 Document Type Definition (DTD) — 173
13.3 Reserved Words — 180

14. A Compact Binary Representation of WML — 180
14.1 Extension Tokens — 180
   14.1.1 Global Extension Tokens — *180*
   14.1.2 Tag Tokens — *180*
   14.1.3 Attribute Tokens — *180*
14.2 Encoding Semantics — 181
   14.2.1 Encoding Variables — *181*
   14.2.2 Encoding Tag and Attributes Names — *181*
   14.2.3 Document Validation — *181*
14.3 Numeric Constants — 181
   14.3.1 WML Extension Token Assignment — *181*
   14.3.2 Tag Tokens — *182*
   14.3.3 Attribute Start Tokens — *183*
   14.3.4 Attribute Value Tokens — *184*
14.4 WML Encoding Examples — 185

15. Static Conformance Statement 186
15.1 WML User Agent 186
15.1.1 Character Set and Encoding 186
15.1.2 Events and Navigation 187
15.1.3 State Model 187
15.1.4 User Agent Semantics 187
15.1.5 Elements 187
15.2 WML Encoder 188
15.2.1 Token Table 188
15.2.2 Validation 189
15.3 WML Document 189

1. Scope

Wireless Application Protocol (WAP) is a result of continuous work to define an industry-wide specification for developing applications that operate over wireless communication networks. The scope for the WAP Forum is to define a set of specifications to be used by service applications. The wireless market is growing very quickly and reaching new customers and services. To enable operators and manufacturers to meet the challenges in advanced services, differentiation and fast/flexible service creation, WAP defines a set of protocols in transport, session and application layers. For additional information on the WAP architecture, refer to *"Wireless Application Protocol Architecture Specification"* [WAP].

This specification defines the Wireless Markup Language (WML). WML is a markup language based on [XML] and is intended for use in specifying content and user interface for narrowband devices, including cellular phones and pagers.

WML is designed with the constraints of small narrowband devices in mind. These constraints include:

- Small display and limited user input facilities
- Narrowband network connection
- Limited memory and computational resources

WML includes four major functional areas:

- Text presentation and layout - WML includes text and image support, including a variety of formatting and layout commands. For example, boldfaced text may be specified.
- Deck/card organisational metaphor - all information in WML is organised into a collection of *cards* and *decks*. Cards specify one or more units of user interaction (e.g., a choice menu, a screen of text or a text entry field). Logically, a user navigates through a series of WML cards, reviews the contents of each, enters requested information, makes choices and moves on to another card.
- Cards are grouped together into decks. A WML deck is similar to an HTML page, in that it is identified by a URL [RFC2396] and is the unit of content transmission.
- Inter-card navigation and linking - WML includes support for explicitly managing the navigation between cards and decks. WML also includes provisions for event handling in the device, which may be used for navigational purposes or to execute scripts. WML also supports anchored links, similar to those found in [HTML4].
- String parameterisation and state management - all WML decks can be parameterised using a state model. Variables can be used in the place of strings and are substituted at runtime. This parameterisation allows for more efficient use of network resources.

2. Document Status

This document is available online in the following formats:
- PDF format at http://www.wapforum.org/.

2.1 Copyright Notice

© Copyright Wireless Application Forum Ltd, 1998, 1999.

Terms and conditions of use are available from the Wireless Application Protocol Forum Ltd. web site at http://www.wapforum.org/docs/copyright.htm.

2.2 Errata

Known problems associated with this document are published at http://www.wapforum.org/.

2.3 Comments

Comments regarding this document can be submitted to the WAP Forum in the manner published at http://www.wapforum.org/.

3. References

3.1 Normative References

[CACHE]   "WAP Caching Model Specification", WAP Forum, 11-Febuary-1999.
            URL: http://www.wapforum.org/

[ISO10646] "Information Technology - Universal Multiple-Octet Coded Character Set (UCS) - Part 1: Architecture and Basic Multilingual Plane", ISO/IEC 10646-1:1993.

[RFC1766]   "Tags for the Identification of Languages", H. Alvestrand, March 1995.
            URL: http://www.ietf.org/rfc/rfc1766.txt

[RFC2045]   "Multipurpose Internet Mail Extensions (MIME) Part One: Format of Internet Message Bodies", N. Freed, et al., November 1996.
            URL: http://www.ietf.org/rfc/rfc2045.txt

[RFC2047]   "MIME (Multipurpose Internet Mail Extensions) Part Three: Message Header Extensions for Non-ASCII Text", K. Moore, November 1996.
            URL: http://www.ietf.org/rfc/rfc2047.txt

[RFC2048]   "Multipurpose Internet Mail Extensions (MIME) Part Four: Registration Procedures", N. Freed, et al., November 1996. URL: http://www.ietf.org/rfc/rfc2048.txt

[RFC2068]   "Hypertext Transfer Protocol - HTTP/1.1", R. Fielding, et al., January 1997.
            URL: http://www.ietf.org/rfc/rfc2068.txt

[RFC2119]   "Key words for use in RFCs to Indicate Requirement Levels", S. Bradner, March 1997.
            URL: http://www.ietf.org/rfc/rfc2119.txt

[RFC2388]   "Returning Values from Forms: multipart/form-data" L. Masinter. August 1998.
            URL: http://www.ietf.org/rfc/rfc2388.txt

[RFC2396]   "Uniform Resource Identifiers (URI): Generic Syntax", T. Berners-Lee, et al., August 1998. URL: http://www.ietf.org/rfc/rfc2396.txt

[UNICODE]   "The Unicode Standard: Version 2.0", The Unicode Consortium, Addison-Wesley Developers Press, 1996. URL: http://www.unicode.org/

[WAE]   "Wireless Application Environment Specification", WAP Forum, 4-November-1999.
            URL: http://www.wapforum.org/

[WAP]   "Wireless Application Protocol Architecture Specification", WAP Forum, 30-April-1998.
            URL: http://www.wapforum.org/
[WBXML]   "Binary XML Content Format Specification", WAP Forum, 4-November-1999.
            URL: http://www.wapforum.org/
[WSP]   "Wireless Session Protocol", WAP Forum, 5-November-1999.
            URL: http://www.wapforum.org/
[XML]   "Extensible Markup Language (XML), W3C Proposed Recommendation
            10-February-1998, REC-xml-19980210", T. Bray, et al, February 10, 1998.
            URL: http://www.w3.org/TR/REC-xml

3.2 Informative References

[HDML2]   "Handheld Device Markup Language Specification", P. King, et al., April 11, 1997.
            URL: http://www.uplanet.com/pub/hdml_w3c/hdml20-1.html
[HTML4]   "HTML 4.0 Specification, W3C Recommendation 18-December-1997, REC-HTML40-
            971218", D. Raggett, et al., September 17, 1997.
            URL: http://www.w3.org/TR/REC-html40
[ISO8879]   "Information Processing - Text and Office Systems - Standard Generalised Markup
            Language (SGML)", ISO 8879:1986.

4. Definitions and Abbreviations

4.1 Definitions

The following are terms and conventions used throughout this specification.

The key words "MUST", "MUST NOT", "REQUIRED", "SHALL", "SHALL NOT", "SHOULD", "SHOULD NOT", "RECOMMENDED", "MAY" and "OPTIONAL" in this document are to be interpreted as described in [RFC2119].

Author - an author is a person or program that writes or generates WML, WMLScript or other content.

Card - a single WML unit of navigation and user interface. May contain information to present to the user, instructions for gathering user input, etc.

Character Encoding - when used as a verb, character encoding refers to conversion between sequence of characters and a sequence of bytes. When used as a noun, character encoding refers to a method for converting a sequence of bytes to a sequence of characters. Typically, WML document character encoding is captured in transport headers attributes (e.g., Content-Type's "charset" parameter), meta information placed within a document, or the XML declaration defined by [XML].

Client - a device (or application) that initiates a request for connection with a server.

Content - subject matter (data) stored or generated at an origin server. Content is typically displayed or interpreted by a user agent in response to a user request.

Content Encoding - when used as a verb, content encoding indicates the act of converting content from one format to another. Typically the resulting format requires less physical space than the original, is easier to process or store and/or is encrypted. When used as a noun, content encoding specifies a particular format or encoding standard or process.

Content Format - actual representation of content.

Deck - a collection of WML cards. A WML deck is also an XML document.

Device - a network entity that is capable of sending and receiving packets of information and has a unique device address. A device can act as both a client or a server within a given context or across multiple contexts. For example, a device can service a number of clients (as a server) while being a client to another server.

JavaScript - a *de facto* standard language that can be used to add dynamic behaviour to HTML documents. JavaScript is one of the originating technologies of ECMAScript.

Man-Machine Interface - a synonym for user interface.

Origin Server - the server on which a given resource resides or is to be created. Often referred to as a web server or an HTTP server.

Resource - a network data object or service that can be identified by a URL. Resources may be available in multiple representations (e.g., multiple languages, data formats, size and resolutions) or vary in other ways.

Server - a device (or application) that passively waits for connection requests from one or more clients. A server may accept or reject a connection request from a client.

SGML - the Standardised Generalised Markup Language (defined in [ISO8879]) is a general-purpose language for domain-specific markup languages.

Terminal - a device providing the user with user agent capabilities, including the ability to request and receive information. Also called a mobile terminal or mobile station.

Transcode - the act of converting from one character set to another, e.g., conversion from UCS-2 to UTF-8.

User - a user is a person who interacts with a user agent to view, hear, or otherwise use a resource.

User Agent - a user agent is any software or device that interprets WML, WMLScript, WTAI or other resources. This may include textual browsers, voice browsers, search engines, etc.

WMLScript - a scripting language used to program the mobile device. WMLScript is an extended subset of the JavaScript scripting language.

XML - the Extensible Markup Language is a World Wide Web Consortium (W3C) standard for Internet markup languages, of which WML is one such language. XML is a restricted subset of SGML.

4.2 Abbreviations

For the purposes of this specification, the following abbreviations apply.

BNF	Backus-Naur Form
HDML	Handheld Markup Language [HDML2]
HTML	HyperText Markup Language [HTML4]
HTTP	HyperText Transfer Protocol [RFC2068]
IANA	Internet Assigned Number Authority
MMI	Man-Machine Interface
PDA	Personal Digital Assistant
RFC	Request For Comments

SGML	Standardised Generalised Markup Language [ISO8879]
UI	User Interface
URL	Uniform Resource Locator [RFC2396]
URN	Uniform Resource Name
W3C	World Wide Web Consortium
WAE	Wireless Application Environment [WAE]
WAP	Wireless Application Protocol [WAP]
WSP	Wireless Session Protocol [WSP]
XML	Extensible Markup Language [XML]

4.3 Device Types

WML is designed to meet the constraints of a wide range of small, narrowband devices. These devices are primarily characterised in four ways:

- **Display size** - smaller screen size and resolution. A small mobile device such as a phone may only have a few lines of textual display, each line containing 8-12 characters.
- **Input devices** - a limited, or special-purpose input device. A phone typically has a numeric keypad and a few additional function-specific keys. A more sophisticated device may have software-programmable buttons, but may not have a mouse or other pointing device.
- **Computational resources** - low power CPU and small memory size; often limited by power constraints.
- **Narrowband network connectivity** - low bandwidth and high latency. Devices with 300bps to 10kbps network connections and 5-10 second round-trip latency are not uncommon.

This document uses the following terms to define broad classes of device functionality:

- **Phone** - the typical display size ranges from two to ten lines. Input is usually accomplished with a combination of a numeric keypad and a few additional function keys. Computational resources and network throughput is typically limited, especially when compared with more general-purpose computer equipment.
- **PDA** - a Personal Digital Assistant is a device with a broader range of capabilities. When used in this document, it specifically refers to devices with additional display and input characteristics. A PDA display often supports resolution in the range of 160x100 pixels. A PDA may support a pointing device, handwriting recognition and a variety of other advanced features.

These terms are meant to define very broad descriptive guidelines and to clarify certain examples in the document.

5. WML and URLs

The World Wide Web is a network of information and devices. Three areas of specification ensure widespread interoperability:

- A unified naming model. Naming is implemented with Uniform Resource Locators (URLs), which provide standard way to name any network resource. See [RFC2396].
- Standard protocols to transport information (e.g., HTTP).
- Standard content types (e.g., HTML, WML).

WML assumes the same reference architecture as HTML and the World Wide Web. Content is named using URLs and is fetched over standard protocols that have HTTP semantics, such as [WSP]. URLs are defined in [RFC2396]. The character set used to specify URLs is also defined in [RFC2396].

In WML, URLs are used in the following situations:

- When specifying navigation, e.g., hyperlinking.
- When specifying external resources, e.g., an image or a script.

5.1 URL Schemes

WML browsers must implement the URL schemes specified in [WAE].

5.2 Fragment Anchors

WML has adopted the HTML *de facto* standard of naming locations within a resource. A WML fragment anchor is specified by the document URL, followed by a hash mark (#), followed by a fragment identifier. WML uses fragment anchors to identify individual WML cards within a WML deck. If no fragment is specified, a URL names an entire deck. In some contexts, the deck URL also implicitly identifies the first card in a deck.

5.3 Relative URLs

WML has adopted the use of relative URLs, as specified in [RFC2396]. [RFC2396] specifies the method used to resolve relative URLs in the context of a WML deck. The base URL of a WML deck is the URL that identifies the deck.

6. WML Character Set

WML is an XML language and inherits the XML document character set. In SGML nomenclature, a document character set is the set of all logical characters that a document type may contain (e.g., the letter 'T' and a fixed integer identifying that letter). An SGML or XML document is simply a sequence of these integer tokens, which taken together form a document.

The document character set for XML and WML is the Universal Character Set of ISO/IEC-10646 ([ISO10646]). Currently, this character set is identical to Unicode 2.0 [UNICODE]. WML will adopt future changes and enhancements to the [XML] and [ISO10646] specifications. Within this document, the terms ISO10646 and Unicode are used interchangeably and indicate the same document character set.

There is no requirement that WML decks be encoded using the full Unicode encoding (e.g., UCS-4). Any character encoding ("charset") that contains a proper subset of the logical characters in Unicode may be used (e.g., US-ASCII, ISO-8859-1, UTF-8, Shift_JIS, etc.). Documents not encoded using UTF-8 or UTF-16 must declare their encoding as specified in the XML specification and should include Content-Type meta-information.

CONFORMANCE RULES:		
WML-01.	UTF-8 Encoding	O
WML-02.	UTF-16 Encoding	O
WML-03.	UCS-4 Encoding	O
WML-04.	Other character encoding	O

6.1 Reference Processing Model

WML documents maybe encoded with any character encoding as defined by [HTML4].

Character encoding of a WML document may be converted to another encoding (or transcoded) to better meet the user agent's characteristics. However, transcoding can lead to loss of information and must be avoided when the user agent supports the document's original encoding. Unnecessary transcoding must be avoided when information loss will result. If required, transcoding should be done before the document is delivered to the user agent.

This specification does not mandate which character encoding a user agent must support.

Since WML is an XML application, the character encoding of a WML document is determined as defined in the XML specification [XML]. In normal cases it is always possible to detect the character encoding of the document (all other cases are error situations). The meta http-equiv statement, if any is present in the document, is never used to determine the character encoding.

If a WML document is transformed into a different format than XML - for example, into the binary WBXML format - then, the rules relevant for that format are used to determine the character encoding.

When an WML document is accompanied by external information (e.g. HTTP or MIME) there may be multiple sources of information available to determine the character encoding. In this case, their relative priority and the preferred method of handling conflict should be specified as part of the higher-level protocol. See, for example, the documentation of the "`text/vnd.wap.wml`" and "`application/vnd.wap.wmlc`" MIME media types.

The WML reference-processing model is as follows. User agents must implement this processing model, or a model that is indistinguishable from it.

- User agents must correctly map to Unicode all characters in any character encoding that they recognise, or they must behave as if they did.
- Any processing of entities is done in the document character set.

A given implementation may choose any internal representation (or representations) that is convenient.

CONFORMANCE RULES:

WML-05.	Reference processing	M

6.2 Character Entities

A given character encoding may not be able to express all characters of the document character set. For such encoding, or when the device characteristics do not allow users to input some document characters directly, authors and users may use character entities (i.e., [XML] character references). Character entities are a character encoding-independent mechanism for entering any character from the document character set.

WML supports both named and numeric character entities. An important consequence of the reference processing model is that all numeric character entities are referenced with respect to the document character set (Unicode) and not to the current document encoding (charset).

This means that `&#302;` always refers to the same logical character, independent of the current character encoding.

WML supports the following character entity formats:

- Named character entities, such as `&amp;` and `&lt;`
- Decimal numeric character entities, such as `&#32;`

- Hexadecimal numeric character entities, such as `&#x20;`

Seven named character entities are particularly important in the processing of WML:

```
<!ENTITY quot    "&#34;">        <!-- quotation mark -->
<!ENTITY amp     "&#38;#38;">    <!-- ampersand -->
<!ENTITY apos    "&#39;">        <!-- apostrophe -->
<!ENTITY lt      "&#38;#60;">    <!-- less than -->
<!ENTITY gt      "&#62;">        <!-- greater than -->
<!ENTITY nbsp    " ">       <!-- non-breaking space -->
<!ENTITY shy     "&#173;">       <!-- soft hyphen (discretionary hyphen) -->
```

CONFORMANCE RULES:

WML-06.   Character entities   M

7. WML Syntax

WML inherits most of its syntactic constructs from XML. Refer to [XML] for in-depth information on syntactical issues.

7.1 Entities

WML text can contain numeric or named character entities. These entities specify specific characters in the document character set. Entities are used to specify characters in the document character set either which must be escaped in WML or which may be difficult to enter in a text editor. For example, the ampersand (&) is represented by the named entity `&amp;`. All entities begin with an ampersand and end with a semicolon.

WML is an XML language. This implies that the ampersand and less-than characters must be escaped when they are used in textual data, i.e., these characters may appear in their literal form only when used as markup delimiters, within a comment, etc. See [XML] for more details.

7.2 Elements

Elements specify all markup and structural information about a WML deck. Elements may contain a start tag, content and an end tag. Elements have one of two structures:

```
<tag> content </tag>
```

or

```
<tag/>
```

Elements containing content are identified by a start tag (`<tag>`) and an end tag (`</tag>`). An empty-element tag (`<tag/>`) identifies elements with no content.

7.3 Attributes

WML attributes specify additional information about an element. More specifically, attributes specify information about an element that is not part of the element's content. Attributes are always specified in the start tag of an element. For example,

```
<tag attr="abcd"/>
```

Attribute names are an XML NAME and are case sensitive.

XML requires that all attribute values be quoted using either double quotation marks (") or single quotation marks ('). Single quote marks can be included within the attribute value when the value is delimited by double quote marks and vice versa. Character entities may be included in an attribute value.

7.4 Comments

WML comments follow the XML commenting style and have the following syntax:

```
<!-- a comment -->
```

Comments are intended for use by the WML author and should not be displayed by the user agent. WML comments cannot be nested.

7.5 Variables

WML cards and decks can be parameterised using variables. To substitute a variable into a card or deck, the following syntax is used:

```
$identifier
$(identifier)
$(identifier:conversion)
```

Parentheses are required if white space does not indicate the end of a variable. Variable syntax has the highest priority in WML, i.e., anywhere the variable syntax is legal, an unescaped '$' character indicates a variable substitution. Variable references are legal in any PCDATA and in any attribute value identified by the vdata entity type (see section 0). Variable references are illegal in attribute values of type CDATA (see section 0). Since XML does not allow for dollar sign characters in other attribute types (for example, ID and NMTOKEN), variable references are also illegal in those attributes

A sequence of two dollar signs ($$) represents a single dollar sign character in all CDATA attribute values and in all #PCDATA text.

See section 0 for more information on variable syntax and semantics.

CONFORMANCE RULES:

WML-59. Variable references may only occur in **vdata** attribute values   M

7.6 Case Sensitivity

XML is a case-sensitive language; WML inherits this characteristic. No case folding is performed when parsing a WML deck. This implies that all WML tags and attributes are case sensitive. In addition, any enumerated attribute values are case sensitive.

7.7 CDATA Section

CDATA sections are used to escape blocks of text and are legal in any PCDATA, e.g., inside an element. CDATA sections begin with the string "<![CDATA[" and end with the string "]]>". For example:

```
<![CDATA[ this is <B> a test ]]>
```

Any text inside a `CDATA` section is treated as literal text and will not be parsed for markup. `CDATA` sections are useful anywhere literal text is convenient.

Refer to the [XML] specification for more information on `CDATA` sections.

7.8 Processing Instructions

WML makes no use of XML processing instructions beyond those explicitly defined in the XML specification.

7.9 Errors

The [XML] specification defines the concept of a **well-formed** XML document. WML decks that violate the definition of a well-formed document are in error. See section 0 for related information.

8. Core WML Data Types

8.1 Character Data

All character data in WML is defined in terms of XML data types. In summary:

- `CDATA` - text which may contain numeric or named character entities. `CDATA` is used only in attribute values.
- `PCDATA` - text which may contain numeric or named character entities. This text may contain tags (`PCDATA` is "Parsed CDATA"). `PCDATA` is used only in elements.
- `NMTOKEN` - a name token, containing any mixture of name characters, as defined by the XML specification.

See [XML] for more details.

8.2 Length

```
<!ENTITY % length   "CDATA">     <!-- [0-9]+ for pixels or [0-9]+"%" for
                                      percentage length -->
```

The `length` type may either be specified as an integer representing the number of pixels of the canvas (screen, paper) or as a percentage of the available horizontal or vertical space. Thus, the value "50" means fifty pixels. For widths, the value "50%" means half of the available horizontal space (between margins, within a canvas, etc.). For heights, the value "50%" means half of the available vertical space (in the current window, the current canvas, etc.).

The integer value consists of one or more decimal digits ([0-9]) followed by an optional percent character (%). The `length` type is only used in attribute values.

8.3 Vdata

```
<!ENTITY % vdata    "CDATA">     <!-- attribute value possibly containing
                                      variable references -->
```

The `vdata` type represents a string that may contain variable references (see section 0). This type is only used in attribute values.

8.4 Flow

```
<!ENTITY % layout    "br">
<!ENTITY % flow      "%text; | %layout; | img | anchor | a | table">
```

The flow type represents "card-level" information. In general, flow is used anywhere general markup can be included.

8.5 HREF

```
<!ENTITY % HREF    "%vdata;">   <!-- URI, URL or URN designating a hypertext
                                     node. May contain variable references -->
```

The HREF type refers to either a relative or an absolute Uniform Resource Locator [RFC2396]. See section 0 for more information.

8.6 Boolean

```
<!ENTITY % boolean    "(true|false)">
```

The boolean type refers to a logical value of true or false.

8.7 Number

```
<!ENTITY % number    "NMTOKEN">   <!-- a number, with format [0-9]+ -->
```

The number type represents an integer value greater than or equal to zero.

8.8 xml:lang

The xml:lang attribute specifies the natural or formal language of an element or its attributes. The value of the attributes is a language code according to [RFC1766]. See [XML] for details on the syntax and specification of the attribute values. The attribute identifies to the user agent the language used text that may be presented to the user (i.e., an element's content and attribute values). The user agent should perform a best effort to present the data according to the specifics of the language. Nested elements can assume the parent's language or use another. Where an element has both text content and text based attribute values that may be presented to the user, authors must use the same language for both. Variable values that are placed in vdata should match the language of the containing element.

An element's language must be established according to the following precedence (from highest to lowest):

1. Based on the xml:lang attribute specified for the element.
2. Based on the xml:lang attribute specified by the closest parent element.
3. Based on any language information included in the transport and document meta data (see sections 0 and 0 for more detail).
4. Based on user agent default preferences.

8.9 The id and class Attributes

All WML elements have two core attributes: id and class that can be used for such tasks as server-side transformations. The id attribute provides an element a unique name within a single deck. The attribute `class` affiliates an element with one or more classes. Multiple elements can be given the same class name. All elements of a single deck with a common class name are considered part of the same class. Class names are cases sensitive. An element can be part of multiple classes if it has multiple unique class names listed in its `class` attribute. Multiple class names within a single attribute must be separated by white space. Redundant class names as well as insignificant white space between class names may be removed. The WML user agent should ignore these attributes.

8.10 ContentType

```
<!ENTITY % ContentType "%vdata;">   <!-- media type. May contain variable
                                         references -->
```

The ContentType type represents the media type defined in [RFC2045]. See section 0 for more information.

9. Events and Navigation

9.1 Navigation and Event Handling

WML includes navigation and event-handling models. The associated elements allow the author to specify the processing of user agent events. Events may be bound to *tasks* by the author; when an event occurs, the bound task is executed. A variety of tasks may be specified, such as navigation to an author-specified URL. Event bindings are declared by several elements, including do and onevent.

9.2 History

WML includes a simple navigational history model that allows the author to manage backward navigation in a convenient and efficient manner. The user agent history is modelled as a stack of URLs that represent the navigational path the user traversed to arrive at the current card. Three operations may be performed on the history stack:

- **Reset** - the history stack may be reset to a state where it only contains the current card. See the `newcontext` attribute (section 0) for more information.
- **Push** - a new URL is pushed onto the history stack as an effect of navigation to a new card.
- **Pop** - the current card's URL (top of the stack) is popped as a result of backward navigation.

The user agent must implement a navigation history. As each card is accessed via an explicitly specified URL, (e.g., via the `href` attribute in `go` element,) an entry for the card is added to the history stack even if it is identical to the most recent entry. At a minimum, each entry must record the absolute URL of the card, the method (get or post) used to access the card, the value of any postfields, and any request headers. The card content is not stored in the history. Variable references are never stored in the history. Any variable references in the history data must be replaced with the current value of the variables before the entry is added to the stack.

The user agent must return the user to the previous card in the history if a `prev` task is executed (see section 0). The execution of `prev` pops the current card URL from the history stack when a prev task is executed.

If the card is part of deck that was originally fetched using an HTTP post method, and the user agent must re-fetch the deck to establish the card, then the user agent must reissue any post data associated with fetching the original deck. The post data values must be the same values used to fetch the original deck. Refer to section 0 for more information on the semantics of `prev`. See [CACHE] for more information on caching semantics.

CONFORMANCE RULES:

WML-07.   History   M

9.3 The Postfield Element

```
<!ELEMENT postfield EMPTY>
<!ATTLIST postfield
  name        %vdata;        #REQUIRED
  value       %vdata;        #REQUIRED
  %coreattrs;
>
```

The `postfield` element specifies a field name and value for transmission to an origin server during a URL request. The actual encoding of the name and value will depend on the method used to communicate with the origin server.

Refer to section 0 for more information on the use of `postfield` in a go element.

Attributes

name=*vdata*

The `name` attribute specifies the field name.

value=*vdata*

The `value` attribute specifies the field value.

Attributes defined elsewhere

id   (see section 0)
class   (see section 0)

CONFORMANCE RULES:

WML-37.   postfield   M

9.4 The Setvar Element

```
<!ELEMENT setvar EMPTY>
<!ATTLIST setvar
  name        %vdata;        #REQUIRED
  value       %vdata;        #REQUIRED
  %coreattrs;
>
```

The `setvar` element specifies the variable to set in the current browser context as a side effect of executing a task. The element must be ignored if the `name` attribute doe not evaluate to a legal variable name at runtime (see section 0). See section 0 for more information on setting variables.

Attributes

name=*vdata*

The `name` attribute specifies the variable name.

    `value=`*`vdata`*

The `value` attribute specifies the value to be assigned to the variable.

Attributes defined elsewhere
    `id` (see section 0)
    `class` (see section 0)

CONFORMANCE RULES:

WML-53.	setvar	M

9.5 Tasks

```
<!ENTITY % task   "go | prev | noop | refresh">
```

Tasks specify processing that is performed in response to an event such as timer expiring, entering a card, or activating an anchor element.

9.5.1 The Go Element

```
<!ELEMENT go (postfield | setvar)*>
<!ATTLIST go
  href            %HREF;              #REQUIRED
  sendreferer     %boolean;           "false"
  method          (post|get)          "get"
  enctype         %ContentType;       "application/x-www-form-urlencoded"
  accept-charset  CDATA               #IMPLIED
  %coreattrs;
>
```

The go element declares a `go` task, indicating navigation to a URI. If the URI names a WML card or deck, it is displayed. A `go` executes a "push" operation on the history stack (see section 0). The UA must ignore all `postfield` elements of a `go` element if the target of the `go` element is a card contained within the current deck.

Refer to section 0 for more information on the semantics of `go`.

Attributes
    `href=`*`HREF`*

The `href` attribute specifies the destination URI, e.g., the URI of the card to display.

    `sendreferer=`*`boolean`*

If this attribute is true, the user agent must specify, for the server's benefit, the URI of the deck containing this task (i.e., the referring deck). This allows a server to perform a form of access control on URIs, based on which decks are linking to them. The URI must be the smallest relative URI possible if it can be relative at all. For example, if `sendreferer=true`, an HTTP based user agent shall indicate the URI of the current deck in the HTTP "Referer" request header [RFC2068].

    `method=`*`(post|get)`*

This attribute specifies the HTTP submission method. Currently, the values of `get` and `post` are accepted and cause the user agent to perform an HTTP `GET` or `POST` respectively.

`enctype=`*`ContentType`*

This attribute specifies the content type used to submit the parameter to the server (when the value of method is post). The default value for this attribute is `application/x-www-form-urlencoded`. Currently, only `application/x-www-form-urlencoded` or `multipart/form-data` can be specified.

When the field values in a submitted form may contain characters not in the US-ASCII character set, it is recommended that the post method and `multipart/form-data` [RFC2388] are used.

User agents must explicitly specify content type for each part. If a part corresponds to a `postfield` element, its content type should be `text/plain`. The charset parameter is required when the content contains characters not in the US-ASCII character set.

`accept-charset=`*`cdata`*

This attribute specifies the list of character encodings for data that the origin server must accept when processing input. The value of this attribute is a comma- or space-separated list of character encoding names (`charset`) as specified in [RFC2045] and [RFC2068]. The IANA Character Set registry defines the public registry for charset values. This list is an exclusive-OR list, i.e., the server must accept any one of the acceptable character encodings.

If the `accept-charset` attribute is not specified or is the reserved string `unknown`, user agents should use the character encoding that was used to transmit the WML deck that contains the `go` element.

Attributes defined elsewhere
  `id` (see section 0)
  `class` (see section 0)

The `go` element may contain one or more `postfield` elements. These elements specify information to be submitted to the origin server during the request. The submission of field data is performed in the following manner:

1. The field name/value pairs are identified and all variables are substituted.
2. The user agent should transcode the field names and values to the correct character set, as specified explicitly by the `accept-charset` or implicitly by the document encoding.
3. If the `href` attribute value is an HTTP URI, the request is performed according to the `method` and `enctype` attribute's value:

METHOD	ENCTYPE	PROCESS
get	application/x-www-form-urlencoded	The field names and values are escaped using URI-escaping and assembled into an application/x-www-form-urlencoded content type. The submission data is appended to the query component of the URI. The result must be a valid query component with the original query part and the postfields combined. An HTTP GET operation is performed on the resulting URL.
	multipart/form-data	Error.

METHOD	ENCTYPE	PROCESS
post	`application/x-www-form-urlencoded`	The field names and values are escaped using URI-escaping and assembled into an `application/x-www-form-urlencoded` content type.
		The submission data is sent as the body of the HTTP POST request.
		The `Content-Type` header must include the charset parameter to indicate the character encoding.
	`multipart/form-data`	The field names and values are encoded as a `multipart/form-data` content type as defined in [RFC2388].

The `Content-Type` header must include the charset parameter to indicate the character encoding when the part contains characters not in the US-ASCII character set.

The submission data is sent as the body of the HTTP POST request.

When `enctype` attribute's value is `application/x-www-form-urlencoded`, the field names and values must be encoded as follows:

1. The field names and values are escaped using URI-escaping, and listed in the order in which the postfields are presented.
2. The name is separated from the value by '=' and name/value pairs are separated from each other by '&'. See [RFC2396] for more information on the URI syntax and its escape sequence.

It is recommended that user agents submit data with an `application/vnd.wap.multipart.form-data` content type when `enctype` attribute has a value of `multipart/form-data`. Some user agents may only support data submission as `application/x-www-form-urlencoded` content type. Such user agents may ignore `enctype` attribute. Thus, it is recommended that an origin server expect either form of submission (i.e., `multipart/form-data` or `application/x-www-form-urlencoded`) when the `enctype` value is `multipart/form-data`. However, the origin server may assume that it will only receive an `application/x-www-form-urlencoded` submission when the `enctype` value is `application/x-www-form-urlencoded`.

For example, the following go element would cause an HTTP GET request to the URL "/foo?x=1":

```
<go href="/foo">
   <postfield name="x" value="1"/>
</go>
```

The following example will cause an HTTP POST to the URL "/bar" with a message entity containing "w=12&y=test":

```
<go href="/bar" method="post">
   <postfield name="w" value="12"/>
   <postfield name="y" value="test"/>
</go>
```

CONFORMANCE RULES:

WML-29.   go   M

9.5.2 The Prev Element

```
<!ELEMENT prev (setvar)*>
<!ATTLIST prev
  %coreattrs;
  >
```

The `prev` element declares a `prev` task, indicating navigation to the previous URI on the history stack. A `prev` performs a "pop" operation on the history stack (see section 0).

Refer to section 0 for more information on the semantics of `prev`.

Attributes defined elsewhere
    id (see section 0)
    class (see section 0)

CONFORMANCE RULES:

WML-39.   prev   M

9.5.3 The Refresh Element

```
<!ELEMENT refresh (setvar)*>
<!ATTLIST refresh
  %coreattrs;
  >
```

The `refresh` element declares a `refresh` task, indicating an update of the user agent context as specified by the `setvar` elements. User-visible side effects of the state changes (e.g., a change in the screen display) occur during the processing of the `refresh` task. A refresh and its side effects must occur even if the elements have no `setvar` elements given that context may change by other means (e.g., `timer`).

Refer to section 0 for more information on the semantics of `refresh`.

Attributes defined elsewhere
    id (see section 0)
    class (see section 0)

CONFORMANCE RULES:

WML-43.   refresh   M

9.5.4 The Noop Element

```
<!ELEMENT noop EMPTY>
<!ATTLIST noop
  %coreattrs;
  >
```

This `noop` element specifies that nothing should be done, i.e., "no operation".

Refer to section 0 for more information on the semantics of `noop`.

Attributes defined elsewhere
>   id (see section 0)
>   class (see section 0)

CONFORMANCE RULES:

WML-35.   noop   M

9.6 Card/Deck Task Shadowing

A variety of elements can be used to create an event binding for a card. These bindings may also be declared at the deck level:

- **Card-level** - the event-handling element may appear inside a card element and specify event-processing behaviour for that particular card.
- **Deck-level** - the event-handling element may appear inside a template element and specify event-processing behaviour for all cards in the deck. A deck-level event-handling element is equivalent to specifying the event-handling element in each card.

A card-level event-handling element overrides (or "shadows") a deck-level event-handling element if they both specify the same event. A card-level onevent element will shadow a deck-level onevent element if they both have the same type. A card-level do element will shadow a deck-level do element if they have the same name.

For a given card, the *active* event-handling elements are defined as the event-handling elements specified in the card that do not bind a noop task, plus any event-handling elements specified in the deck's template not overridden (or shadowed) in the card or bind a noop task. Shadowed event handling elements, event-handling elements defined in other cards, and event-handling elements that bind a noop task are considered *inactive* elements.

If a card-level element shadows a deck-level element and the card-level element binds a noop task, the event binding for that event will be completely masked. In this situation, the card- and deck-level elements will be ignored and no side effects will occur on delivery of the event. In other words, both the card-level and the deck-level elements are considered inactive elements in such a case.

If a card-level element or deck-level element binds a noop task but does not shadow and is not shadowed by another element, then the binding for that event will also be masked and similarly ignored with no side effects.

In the following example, a deck-level do element indicates that a prev task should execute on receipt of a particular user action. The first card inherits the do element specified in the template element and will display the do to the user. The second card shadows the deck-level do element with a noop. The user agent will not display the do element when displaying the second card. The third card shadows the deck-level do element, causing the user agent to display the alternative label and to perform the go task if the do is selected.

```
<wml>
    <template>
        <do type="options" name="do1" label="default">
<prev/>
</do>
    </template>

    <card id="first">
        <!-- deck-level do not shadowed.  The card exposes the
```

```
                deck-level do as part of the current card -->

        <!-- rest of card -->
        …
    </card>

    <card id="second">
<!-- deck-level do is shadowed with noop.
    It is not exposed to the user -->
        <do type="options" name="do1">
<noop/>
</do>

        <!-- rest of card -->
        …
    </card>

    <card id="third">
<!-- deck-level do is shadowed. It is replaced by a card-level do -->
        <do type="options" name="do1" label="options">
            <go href="/options"/>
        </do>

        <!-- rest of card -->
        …
    </card>
</wml>
```

CONFORMANCE RULES:

WML-08.   Card/Deck task Shadowing   M

9.7 The Do Element

```
<!ENTITY % task    "go | prev | noop | refresh">
<!ELEMENT do (%task;)>
<!ATTLIST do
    type        CDATA           #REQUIRED
    label       %vdata;         #IMPLIED
    name        NMTOKEN         #IMPLIED
    optional    %boolean;       "false"
    xml:lang    NMTOKEN         #IMPLIED
    %coreattrs;
>
```

The do element provides a general mechanism for the user to act upon the current card, i.e., a card-level user interface element. The representation of the do element is user agent dependent and the author must only assume that the element is mapped to a unique user interface *widget* that the user can activate. For example, the widget mapping may be to a graphically rendered button, a soft or function key, a voice-activated command sequence, or any other interface that has a simple "activate" operation with no inter-operation persistent state. When the user activates a do element, the associated task is executed.

The do element may appear at both the card and deck-level:

- **Card-level** - the do element may appear inside a `card` element and may be located anywhere in the text flow. If the user agent intends to render the do element inline (i.e., in the text flow), it should use the element's anchor point as the rendering point. WML authors must not rely on the inline rendering of the do element and must not rely on the correct positioning of an inline rendering of the element.
- **Deck-level** - the do element may appear inside a `template` element, indicating a deck-level do element. A deck-level do element applies to all cards in the deck (i.e., is equivalent to having specified the do within each card). For the purposes of inline rendering, the user agent must behave as if deck-level do elements are located at the end of the card's text flow.

Attributes

`type=cdata`

The do element type. This attribute provides a hint to the user agent about the author's intended use of the element and should be used by the user agent to provide a suitable mapping onto a physical user interface construct. WML authors must not rely on the semantics or behaviour of an individual `type` value, or on the mapping of `type` to a particular physical construct. All types are reserved, except for those marked as experimental, or vendor-specific.

User agents must accept any `type`, but may treat any unrecognised type as the equivalent of `unknown`.

In the following table, the * character represents any string, e.g., `Test*` indicates any string starting with the word `Test`. Although experimental and vendor-specific types may be specified in any case, they are case-sensitive; e.g., the types `VND-foo` and `vnd-foo` are distinct.

Table 1. Pre-defined DO types

TYPE	DESCRIPTION
accept	Positive acknowledgement (acceptance)
prev	Backward history navigation
help	Request for help. May be context-sensitive.
reset	Clearing or resetting state.
options	Context-sensitive request for options or additional operations.
delete	Delete item or choice.
unknown	A generic do element. Equivalent to an empty string (e.g., `type=" "`).
X-*, x-*	Experimental types. This set is not reserved.
vnd.*, VND.* and any combination of [Vv][Nn][Dd].*	Vendor-specific or user-agent-specific types. This set is not reserved. Vendors should allocate names with the format VND.CO-TYPE, where CO is a company name abbreviation and `type` is the do element type. See [RFC2045] for more information.

`label=vdata`

If the user agent is able to dynamically label the user interface widget, this attribute specifies a textual string suitable for such labelling. The user agent must make a best-effort attempt to label the UI widget and should adapt the label to the constraints of the widget (e.g., truncate the string). If an element can not be dynamically labelled, this attribute may be ignored.

To work well on a variety of user agents, labels should be six characters or shorter in length.

>   name=nmtoken

This attribute specifies the name of the do event binding. If two do elements are specified with the same name, they refer to the same binding. A card-level do element shadows a deck-level do with the same name value (see section 9.6 for more detail). It is an error to specify two or more do elements with the same name in a single card or in the template element. An unspecified name defaults to the value of the type attribute.

>   optional=boolean

If this attribute has a value of true, the user agent may ignore this element.

All active do elements that have not been designated optional must be presented to the user. The UA must make such elements accessible to the user in some manner. In other words, it must be possible for the user to activate them (i.e., initiate the tasks associated with the element), via some user interface element.

All inactive do elements must not be presented to the user in a way that the user can activate such elements. Elements that have been designated as optional may be ignored at the discretion of the user agent. (For example, the author may wish to allow the user agent to ignore vendor-specific types that it doesn't recognise.)

Attributes defined elsewhere
>   xml:lang (see section 0)
>   id (see section 0)
>   class (see section 0)

CONFORMANCE RULES:

WML-26.	do	M

9.8 The Anchor Element

```
<!ELEMENT anchor ( #PCDATA | br | img | go | prev | refresh )*>
<!ATTLIST anchor
    title       %vdata;     #IMPLIED
    accesskey   %vdata;     #IMPLIED
    xml:lang    NMTOKEN     #IMPLIED
    %coreattrs;
    >
```

The anchor element specifies the head of a link. The tail of a link is specified as part of other elements (e.g., a card name attribute). It is an error to nest anchored links.

Anchors may be present in any text flow, excluding the text in option elements (i.e., anywhere formatted text is legal, except for option elements). Anchored links have an associated *task* that specifies the behaviour when the anchor is selected. It is an error to specify other than one task element (e.g., go, prev or refresh) in an anchor element.

Attributes
>   title=vdata

This attribute specifies a brief text string identifying the link. The user agent may display it in a variety of ways, including dynamic labelling of a button or key, a *tool tip*, a voice prompt, etc. The user agent may truncate or ignore this attribute depending on the characteristics of the navigational user interface. To work well on a broad range of user agents, the author should limit all labels to 6 characters in length.

```
accesskey=vdata
```

This attribute assigns an access key to an element. An access key is a single character from the document character set. Its purpose is to allow the user to activate a particular element by using a single key. The keys available will vary depending on the type of mobile device being used (e.g., phones will usually have "0"-"9", "*" and "#" keys).

The user agent is not required to support `accesskey`. If the user agent does supports access keys, it should respect the requested value if possible. When this is not possible (e.g., the requested key does not already exist or has been defined more than once,) the user agent should assign available keys to the remaining elements that request them, in the order they are encountered in the card, until all available keys are assigned. Any remaining `accesskey` attributes are ignored. The author can not assume that the key specified will be the one used, nor that all elements that define an `accesskey` will have one assigned.

The following elements support the accesskey attribute: A, INPUT, and ANCHOR.

Activating an access key assigned to an element gives focus to the element. The action that occurs when an element receives focus depends on the element. For example, when a user activates a link defined by the A element, the user agent generally follows the link. When a user activates a radio button, the user agent changes the value of the radio button. When the user activates a text field, it allows input, etc.

In this example, we request an access key for a link defined by the A element. Activating this access key takes the user to another document, in this case, a table of contents.

```
<a accesskey="1" href="http://someplace.com/specification/contents.html">
Table of Contents</a>
```

The invocation of access keys may depend on both the user agent and the device on which it is running. For example, the access key may be recognised directly, or it may be necessary to press it in conjunction with a "command" key. The recognition of access keys may be inhibited by context. For example, a device that recognises access keys directly may inhibit their recognition when an input element is active

The rendering of access keys depends on the user agent. User agents should render the value of an access key in such a way as to emphasise its role and to distinguish it from other characters (e.g., by underlining it, enclosing it in brackets, displaying it in reverse video or in a distinctive font, etc.) The author should not refer to the specific access key value in content or in documentation, as the requested value may not be used.

Attributes defined elsewhere

```
xml:lang  (see section 0)
id  (see section 0)
class  (see section 0)
```

CONFORMANCE RULES:

WML-20.	anchor	M

9.9 The A Element

```
<!ELEMENT a ( #PCDATA | br | img )*>
<!ATTLIST a
```

```
href          %HREF;      #REQUIRED
title         %vdata;     #IMPLIED
accesskey     %vdata;     #IMPLIED
xml:lang      NMTOKEN     #IMPLIED
%coreattrs;
>
```

The `a` element is a short form of the `anchor` element, and is bound to a go task without variables. For example, the following markup:

```
<anchor>follow me
<go href="destination"/>
</anchor>
```

Is identical in behaviour and semantics to:

```
<a href="destination">follow me</a>
```

It is invalid to nest a elements. Authors are encouraged to use the `a` element instead of anchor where possible, to allow more efficient tokenisation.

Attributes defined elsewhere
xml:lang (see section 0)
id (see section 0)
class (see section 0)
accesskey (see section 0)

CONFORMANCE RULES:

WML-19.   a   M

9.10 Intrinsic Events

Several WML elements are capable of generating events when the user interacts with them. These so-called "intrinsic events" indicate state transitions inside the user agent. Individual elements specify the events they can generate. WML defines the following intrinsic events:

Table 2. WML Intrinsic Events

EVENT	ELEMENT(S)	DESCRIPTION
ontimer	card, template	The `ontimer` event occurs when a timer expires. Timers are specified using the `timer` element (see section 0).
onenterforward	card, template	The `onenterforward` event occurs when the user causes the user agent to enter a card using a go task or any method with identical semantics. This includes card entry caused by a script function or user-agent-specific mechanisms, such as a means to directly enter and navigate to a URL.
		The `onenterforward` intrinsic event may be specified at both the card and deck-level. Event bindings specified in the `template` element apply to all cards in the deck and may be overridden as specified in section 0.

EVENT	ELEMENT(S)	DESCRIPTION
onenterbackward	card, template	The onenterbackward event occurs when the user causes the user agent to navigate into a card using a prev task or any method with identical semantics. In other words, the onenterbackward event occurs when the user causes the user agent to navigate into a card by using a URL retrieved from the history stack. This includes navigation caused by a script function or user-agent-specific mechanisms.
		The onenterbackward intrinsic event may be specified at both the card and deck-level. Event bindings specified in the template element apply to all cards in the deck and may be overridden as specified in section 0.
onpick	option	The onpick event occurs when the user selects or deselects this item.

The author may specify that certain tasks are to be executed when an intrinsic event occurs. This specification may take one of two forms. The first form specifies a URI to be navigated to when the event occurs. This event binding is specified in a well-defined element-specific attribute and is the equivalent of a go task. For example:

```
<card onenterforward="/url"> <p> Hello </p> </card>
```

This attribute value may only specify a URL.

The second form is an expanded version of the previous, allowing the author more control over user agent behaviour. An onevent element is declared within a parent element, specifying the full event binding for a particular intrinsic event. For example, the following is identical to the previous example:

```
<card>
    <onevent type="onenterforward">
<go href="/url"/>
    </onevent>
    <p>
   Hello
    </p>
</card>
```

The user agent must treat the attribute syntax as an abbreviated form of the onevent element where the attribute name is mapped to the onevent type.

An intrinsic event binding is scoped to the element in which it is declared, e.g., an event binding declared in a card is local to that card. Any event binding declared in an element is active only within that element. Event bindings specified in sub-elements take precedence over any conflicting event bindings declared in a parent element. Conflicting event bindings within an element are an error.

CONFORMANCE RULES:

| WML-09. | Intrinsic Events | M |
| WML-64. | Event bindings must not conflict | M |

9.10.1 The Onevent Element

```
<!ENTITY % task    "go | prev | noop | refresh">
<!ELEMENT onevent (%task;)>
<!ATTLIST onevent
  type          CDATA       #REQUIRED
  %coreattrs;
  >
```

The onevent element binds a task to a particular intrinsic event for the immediately enclosing element, i.e., specifying an onevent element inside an "XYZ" element associates an intrinsic event binding with the "XYZ" element.

The user agent must ignore any onevent element specifying a type that does not correspond to a legal intrinsic event for the immediately enclosing element.

Attributes
   type=cdata

The type attribute indicates the name of the intrinsic event.

Attributes defined elsewhere
   id (see section 0)
   class (see section 0)

CONFORMANCE RULES:

WML-40.   onevent   M

9.10.2 Card/Deck Intrinsic Events

The onenterforward and onenterbackward intrinsic events may be specified at both the card- and deck-level and have the shadowing semantics defined in section 0. Intrinsic events may be overridden regardless of the syntax used to specify them. A deck-level event-handler specified with the onevent element may be overridden by the onenterforward attribute and vice versa.

10. The State Model

WML includes support for managing user agent state, including:

- **Variables** - parameters used to change the characteristics and content of a WML card or deck;
- **History** - navigational history, which may be used to facilitate efficient backward navigation; and
- **Implementation-dependent state** - other state relating to the particulars of the user agent implementation and behaviour.

10.1 The Browser Context

WML state is stored in a single scope, known as a *browser context*. The browser context is used to manage all parameters and user agent state, including variables, the navigation history and other implementation-dependent information related to the current state of the user agent.

CONFORMANCE RULES:

WML-10.   Browser context   M

10.2 The Newcontext Attribute

The browser context may be initialised to a well-defined state by the `newcontext` attribute of the `card` element (see section 0). This attribute indicates that the browser context should be re-initialised and must perform the following operations:

- Unset (remove) all variables defined in the current browser context,
- Clear the navigational history state, and
- Reset implementation-specific state to a well-known value.

The `newcontext` is only performed as part of the `go` task. See section 0 for more information on the processing of state during navigation.

CONFORMANCE RULES:

WML-11.	Initialisation (newcontext)	M

10.3 Variables

All WML content can be parameterised, allowing the author a great deal of flexibility in creating cards and decks with improved caching behaviour and better perceived interactivity. WML variables can be used in the place of strings and are substituted at run-time with their current value.

A variable is said to be *set* if it has a value not equal to the empty string. A value is *not set* if it has a value equal to the empty string, or is otherwise unknown or undefined in the current browser context.

CONFORMANCE RULES:

WML-12.	Variables	M

10.3.1 Variable Substitution

The values of variables can be substituted into both the text (`#PCDATA`) of a card and into `%vdata` and `%HREF` attribute values in WML elements. Only textual information can be substituted; no substitution of elements or attributes is possible. The substitution of variable values happens at run-time in the user agent. Substitution does not affect the current value of the variable and is defined as a string substitution operation. If an undefined variable is referenced, it results in the substitution of the empty string.

WML variable names consist of an US-ASCII letter or underscore followed by zero or more letters, digits or underscores. Any other characters are illegal and result in an error. Variable names are case sensitive.

The following description of the variable substitution syntax uses the Extended Backus-Naur Form (EBNF) established in [XML].

```
var     = ( "$" varname ) |
          ( "$(" varname ( conv )? ")" ) |
          ( deprecated-var )

conv    = ":" ( "escape" | "noesc" | "unesc" )

varname = ( "_" | alpha ) ( "_" | alpha | digit )*
alpha   = lalpha | halpha
lalpha  = "a" | "b" | "c" | "d" | "e" | "f" | "g" | "h" | "i" |
          "j" | "k" | "l" | "m" | "n" | "o" | "p" | "q" | "r" |
```

```
              "s"  |  "t"  |  "u"  |  "v"  |  "w"  |  "x"  |  "y"  |  "z"
halpha   =    "A"  |  "B"  |  "C"  |  "D"  |  "E"  |  "F"  |  "G"  |  "H"  |  "I"  |
              "J"  |  "K"  |  "L"  |  "M"  |  "N"  |  "O"  |  "P"  |  "Q"  |  "R"  |
              "S"  |  "T"  |  "U"  |  "V"  |  "W"  |  "X"  |  "Y"  |  "Z"
digit    =    "0"  |  "1"  |  "2"  |  "3"  |  "4"  |  "5"  |  "6"  |  "7"  |
              "8"  |  "9"

deprecated-var  =  "$(" varname deprecated-conv ")"
deprecated-conv =  ":"( escape | noesc | unesc )
escape          =  ("E" | "e") ( ( "S" | "s" ) ( "C" | "c" )
                                 ( "A" | "a" ) ( "P" | "p" )
                                 ( "E" | "e" ) )?
noesc           =  ( "N" | "n" ) ( ( "O" | "o" ) ("E" | "e")
                                   ( "S" | "s" ) ( "C" | "c" ) )?
unesc           =  ( "U" | "u" ) ( ( "N" | "n" ) ("E" | "e")
                                   ( "S" | "s" ) ( "C" | "c" ) )?
```

Mixed case and abbreviated conversions have been **deprecated**. Authors should only use lower case conversions.

Parentheses are required anywhere the end of a variable cannot be inferred from the surrounding context, e.g., an illegal character such as white space.

For example:

```
This is a $var
This is another $(var).
This is an escaped $(var:escape).
This is an unescape $(var:unesc).
Other legal variable forms: $_X $X32 $Test_9A
```

The value of variables can be converted into a different form as they are substituted. A conversion can be specified in the variable reference following the colon. The following table summarised the current conversions and their legal abbreviations:

Table 3. Variable Escaping Methods

CONVERSION	EFFECT
Noesc	No change to the value of the variable.
Escape	URL-escape the value of the variable.
Unesc	URL-unescape the value of the variable.

The use of a conversion during variable substitution does not affect the actual value of the variable.

URL-escaping is detailed in [RFC2396]. All lexically sensitive characters defined in WML must be escaped, including all characters not in the `unreserved` set specified by [RFC2396].

If no conversion is specified, the variable is substituted using the conversion format appropriate for the context. All attributes defined as %HREF; default to `escape` conversion, elsewhere no conversion is done. Specifying the `noesc` conversion disables context sensitive escaping of a variable.

CONFORMANCE RULES:

WML-60.	Variable references must match the production rule **var**	M

10.3.2 Parsing the Variable Substitution Syntax

The variable substitution syntax (e.g., $X) is parsed after all XML parsing is complete. In XML terminology, variable substitution is parsed after the *XML processor* has parsed the document and provided the resulting parsed form to the *XML application*. In the context of this specification, the WML parser and user agent is the *XML application*.

This implies that all variable syntax is parsed *after* the XML constructs, such as tags and entities, have been parsed. In the context of variable parsing, all XML syntax has a higher precedence than the variable syntax, e.g., entity substitution occurs before the variable substitution syntax is parsed. The following examples are identical references to the variable named X:

```
$X
&#x24;X
$&#x58;
&#36;&#x58;
```

10.3.3 The Dollar-sign Character

A side effect of the parsing rules is that the literal dollar sign must be encoded with a pair of dollar sign entities in any `#PCDATA` text or `CDATA` attribute values. A single dollar-sign entity, even specified as `&#x24;`, results in a variable substitution.

In order to include a '$' character in a WML deck, it must be explicitly escaped. This can be accomplished with the following syntax:

```
$$
```

Two dollar signs in a row are replaced with a single '$' character. For example:

```
This is a $$ character.
```

This would be displayed as:

```
This is a $ character.
```

To include the '$' character in URL-escaped strings, specify it with the URL-escaped form:

```
%24
```

10.3.4 Setting Variables

There are a number of ways to set the value of a variable. When a variable is set and it is already defined in the browser context, the current value is updated.

The `setvar` element allows the author to set variable state as a side effect of navigation. `Setvar` may be specified in task elements, including `go`, `prev` and `refresh`. The `setvar` element specifies a variable name and value, for example:

```
<setvar name="location" value="$(X)"/>
```

The variable specified in the `name` attribute (e.g., `location`) is set as a side effect of navigation. See the discussion of event handling (section 0 and section 0) for more information on the processing of the `setvar` element.

Input elements set the variable identified by the `name` attribute to any information entered by the user. For example, an `input` element assigns the entered text to the variable, and the `select` element assigns the value present in the `value` attribute of the chosen `option` element.

User input is written to variables when the user commits the input to the `input` or `select` element. Committing input is an MMI dependent concept, and the WML author must not rely on a particular user interface. For example, some implementations will update the variable with each character entered into an `input` element, and others will defer the variable update until the `input` element has lost focus. The user agent must update all variables prior to the execution of any task. The user agent may re-display the current card when variables are set, but the author must not assume that this action will occur.

10.3.5 Validation

Within the WML document, any string following a single dollar sign ('$') must be treated as a variable reference and validated, unless it is part of an escaped literal dollar sign sequence according to section 0. Each reference must use proper variable name syntax, according to section 0. Each reference must be placed either within a card's text (#PCDATA) or within %vdata or %HREF attribute values. Other CDATA attribute values must use escaped literal dollar sign as required to prevent the creation of an otherwise valid variable reference. The deck is in error if any variable reference uses invalid syntax or is placed in an invalid location.

Examples of invalid variable use:

```
    <!-- bad variable syntax -->
Balance left is $10.00

<!-- bad placement (in the type attribute) -->
    <do type="x-$(type)" label="$type">
```

Example of escaped dollar sign in an attribute of type CDATA:
```
<!-- Dollar sign escaped in type attribute -->
<do type="x-$$(type)" label="$type">
```

10.4 Context Restrictions

User agents may provide users means to reference and navigate to resources independent of the current content. For example, user agents may provide bookmarks, a URL input dialog, and so forth. Whenever a user agent navigates to a resource that was not the result of an interaction with the content in the current context, the user agent must establish another context for that navigation. The user agent may terminate the current context before establishing another one for the new navigation attempt.

CONFORMANCE RULES:

WML-13.   Context restrictions   M

11. The Structure of WML Decks

WML data are structured as a collection of *cards*. A single collection of cards is referred to as a WML *deck*. Each card contains structured content and navigation specifications. Logically, a user navigates through a series of cards, reviews the contents of each, enters requested information, makes choices and navigates to another card or returns to a previously visited card.

11.1 Document Prologue

A valid WML deck is a valid XML document and therefore must contain an XML declaration and a document type declaration (see [XML] for more detail about the definition of a valid document). A typical document prologue contains:

```
<?xml version="1.0"?>
<!DOCTYPE wml PUBLIC "-//WAPFORUM//DTD WML 1.2//EN"
        "http://www.wapforum.org/DTD/wml_1.2.xml">
```

It is an error to omit the prologue.

11.2 The WML Element

```
<!ELEMENT wml ( head?, template?, card+ )>
<!ATTLIST wml
  xml:lang          NMTOKEN        #IMPLIED
  %coreattrs;
  >
```

The `wml` element defines a deck and encloses all information and cards in the deck.

Attributes

xml:lang=nmtoken

The `xml:lang` attribute specifies the natural or formal language in which the document is written. See section 0 for more detail.

CONFORMANCE RULES:

WML-54.	wml	M

11.2.1 A WML Example

The following is a deck containing two cards, each represented by a `card` element (see section 0 for information on cards). After loading the deck, a user agent displays the first card. If the user activates the DO element, the user agent displays the second card.

```
<wml>
   <card>
      <p>
      <do type="accept">
<go href="#card2"/>
      </do>
      Hello world!
This is the first card...
</p>
   </card>

   <card id="card2">
      <p>
      This is the second card.
Goodbye.
</p>
   </card>
</wml>
```

© Copyright Wireless Application Protocol Forum, Ltd., 1998, 1999
All rights reserved

11.3 The Head Element

```
<!ELEMENT head ( access | meta )+>
<!ATTLIST head
  %coreattrs;
  >
```

The `head` element contains information relating to the deck as a whole, including meta-data and access control elements.

Attributes defined elsewhere
    `id` (see section 0)
    `class` (see section 0)

CONFORMANCE RULES:

WML-30.	head	M

11.3.1 The Access Element

```
<!ELEMENT access EMPTY>
<!ATTLIST access
  domain     CDATA     #IMPLIED
  path       CDATA     #IMPLIED
  %coreattrs;
  >
```

The `access` element specifies access control information for the entire deck. It is an error for a deck to contain more than one `access` element. If a deck does not include an `access` element, access control is disabled. When access control is disabled, cards in any deck can access this deck.

Attributes
    `domain=`*cdata*
    `path=`*cdata*

A deck's `domain` and `path` attributes specify which other decks may access it. As the user agent navigates from one deck to another, it performs access control checks to determine whether the destination deck allows access from the current deck.

If a deck has a `domain` and/or `path` attribute, the referring deck's URI must match the values of the attributes. Matching is done as follows: the access domain is suffix-matched against the domain name portion of the referring URI and the access path is prefix matched against the path portion of the referring URI.

Domain suffix matching is done using the entire element of each sub-domain and must match each element exactly (e.g., `www.wapforum.org` shall match `wapforum.org`, but shall not match `forum.org`). Path prefix matching is done using entire path elements and must match each element exactly (e.g., `/X/Y` matches path=`"/X"` attribute, but does not match path=`"/XZ"` attribute).

The `domain` attribute defaults to the current deck's domain. The `path` attribute defaults to the value "/".

To simplify the development of applications that may not know the absolute path to the current deck, the `path` attribute accepts relative URIs. The user agent converts the relative path to an absolute path and then performs prefix matching against the `PATH` attribute.

For example, given the following access control attributes:

```
domain="wapforum.org"
path="/cbb"
```

The following referring URIs would be allowed to go to the deck:

```
http://wapforum.org/cbb/stocks.cgi
https://www.wapforum.org/cbb/bonds.cgi
http://www.wapforum.org/cbb/demos/alpha/packages.cgi?x=123&y=456
```

The following referring URIs would not be allowed to go to the deck:

```
http://www.test.net/cbb
http://www.wapforum.org/internal/foo.wml
```

Domain and path follow URL capitalisation rules.

Attributes defined elsewhere
id (see section 0)
class (see section 0)

CONFORMANCE RULES:

WML-21.   access   M

11.3.2 The Meta Element

```
<!ELEMENT meta EMPTY>
<!ATTLIST meta
  http-equiv    CDATA       #IMPLIED
  name          CDATA       #IMPLIED
  forua         %boolean;   "false"
  content       CDATA       #REQUIRED
  scheme        CDATA       #IMPLIED
  %coreattrs;
>
```

The meta element contains generic meta-information relating to the WML deck. Meta-information is specified with property names and values. This specification does not define any properties, nor does it define how user agents must interpret meta-data. User agents are not required to support the meta-data mechanism.

A meta element must contain exactly one attribute specifying a property name; i.e., exactly one from the following set: http-equiv and name.

Attributes

name=cdata

This attribute specifies the property name. The user agent must ignore any meta-data named with this attribute. Network servers should not emit WML content containing meta-data named with this attribute.

http-equiv=cdata

This attribute may be used in place of name and indicates that the property should be interpreted as an HTTP header (see [RFC2068]). Meta-data named with this attribute should be converted to a WSP or HTTP response header if the content is tokenised before it arrives at the user agent.

```
forua=boolean
```

If the value is "false" an intermediate agent MUST remove the "meta" element before the document is sent to the client. If the value is "true" the meta data of the element MUST be delivered to the user-agent. The method of delivery may vary. For example, `http-equiv` meta-data may be delivered using HTTP or WSP headers.

```
content=cdata
```

This attribute specifies the property value.

```
scheme=cdata
```

This attribute specifies a form or structure that may be used to interpret the property value. Scheme values vary depending on the type of meta-data.

Attributes defined elsewhere
id (see section 0)
class (see section 0)

CONFORMANCE RULES:

WML-34.	meta	O
WML-61.	Two or more **do** elements with the same **name** must not be present in a single card or in the **template** element. (Note: An unspecified **name** defaults to the value of the **type** attribute.)	M
WML-62.	A **meta** element must not contain more than one attribute of **name** and **http-equiv**.	M

11.4 The Template Element

```
<!ENTITY % navelmts "do | onevent">
<!ELEMENT template (%navelmts;)*>
<!ATTLIST template
  %cardev;
  %coreattrs;
>
```

The template element declares a template for cards in the deck. Event bindings specified in the template element (e.g., do or onevent) apply to all cards in the deck. Specifying an event binding in the template element is equivalent to specifying it in every card element. A card element may override the behaviour specified in the template element. In particular:

- DO elements specified in the `template` element may be overridden in individual cards if both elements have the same NAME attribute value. See section 0 for more information.
- Intrinsic event bindings specified in the `template` element may be overridden by the specification of an event binding in a card element. See section 0 for more information.

See section 0 for the definition of the card-level intrinsic events (the `cardev` entity).

Attributes defined elsewhere
id (see section 0)
class (see section 0)

```
onenterforward   (see section 0)
onenterbackward  (see section 0)
ontimer          (see section 0)
```

CONFORMANCE RULES:

WML-48.   template   M

11.5 The Card Element

A WML deck contains a collection of cards. Each card can contain a variety of content. The user's interaction with the card depends on the type of content the card contains and how the content is rendered by the user agent.

CONFORMANCE RULES:

WML-25.   card   M

11.5.1 Card Intrinsic Events

```
<!ENTITY % cardev
 "onenterforward   %HREF;         #IMPLIED
  onenterbackward  %HREF;         #IMPLIED
  ontimer          %HREF;         #IMPLIED"
 >
```

The following attributes are available in the `card` and `template` elements.

Attributes

onenterforward=HREF

The `onenterforward` event occurs when the user causes the user agent to navigate into a card using a `go` task.

onenterbackward=HREF

The `onenterbackward` event occurs when the user causes the user agent to navigate into a card using a `prev` task.

ontimer=HREF

The `ontimer` event occurs when a `timer` expires.

11.5.2 The Card Element

```
<!ELEMENT card (onevent*, timer?, (do | p | pre)*)>
<!ATTLIST card
  title         %vdata;        #IMPLIED
  newcontext    %boolean;      "false"
  ordered       %boolean;      "true"
  xml:lang      NMTOKEN        #IMPLIED
  %cardev;
  %coreattrs;
  >
```

The `card` element is a container of text and input elements that is sufficiently flexible to allow presentation and layout in a wide variety of devices, with a wide variety of display and input characteristics. The `card` element indicates the general layout and required input fields, but does not overly constrain the user agent implementation in the areas of layout or user input. For example, a `card` can be presented as a single page on a large-screen device and as a series of smaller pages on a small-screen device.

A `card` can contain markup, input fields and elements indicating the structure of the card. The order of elements in the card is significant and should be respected by the user agent. A card's id may be used as a fragment anchor. See section 0 for more information.

Attributes

>   `title=`*`vdata`*

The `title` attribute specifies advisory information about the card. The title may be rendered in a variety of ways by the user agent (e.g., suggested bookmark name, pop-up *tooltip*, etc.).

>   `newcontext=`*`boolean`*

This attribute indicates that the current browser context should be re-initialised upon entry to this card. See section 0 for more information.

>   `ordered=`*`boolean`*

This attribute specifies a hint to the user agent about the organisation of the `card` content. This hint may be used to organise the content presentation or to otherwise influence layout of the card.

- `ordered="true"` - the card is naturally organised as a linear sequence of field elements, e.g., a set of questions or fields which are naturally handled by the user in the order in which they are specified in the group. This style is best for short forms in which no fields are optional (e.g., sending an email message requires a `To:` address, a subject and a message, and they are logically specified in this order).
- It is expected that in small-screen devices, `ordered` groups may be presented as a sequence of screens, with a screen flip in between each field or fieldset. Other user agents may elect to present all fields simultaneously.
- `ordered="false"` - the card is a collection of field elements without a natural order. This is useful for collections of fields containing optional or unordered components or simple record data where the user is updating individual input fields.

It is expected that in small-screen devices, unordered groups may be presented by using a hierarchical or tree organisation. In these types of presentation, the `title` attribute of each field and fieldset may be used to define the name presented to the user in the top-level summary card. A user agent may organise an unordered collection of elements in an ordered fashion.

The user agent may interpret the ordered attribute in a manner appropriate to its device capabilities (e.g., screen size or input device). In addition, the user agent should adopt user interface conventions for handling the editing of input elements in a manner that best suits the device's input model.

For example, a phone-class device displaying a `card` with `ordered="false"` may use a softkey or button to select individual fields for editing or viewing. A PDA-class device might create soft buttons on demand, or simply present all fields on the screen for direct manipulation.

On devices with limited display capabilities, it is often necessary to insert screen flips or other user-interface transitions between fields. When this is done, the user agent needs to decide on the proper boundary between fields. User agents may use the following heuristic for determining the choice of a screen flip location:

fieldset defines a logical boundary between fields.

- Fields (e.g., `input`) may be individually displayed. When this is done, the line of markup (`flow`) immediately preceding the field should be treated as a field prompt and displayed with the input element. The `table` must be treated differently than `input` and `select`. The user agent must insert a line break before each `table` element, except when it is the first non-whitespace markup in a card. The user agent must insert a line break after each `table` element, except when it is the final element in a card.

Attributes defined elsewhere

`xml:lang` (see section 0)
`id` (see section 0)
`class` (see section 0)
`onenterforward` (see section 0)
`onenterbackward` (see section 0)
`ontimer` (see section 0)

11.5.2.1 A Card Example

The following is an example of a simple `card` element embedded within a WML deck. The card contains text, which is displayed by the user agent. In addition, the example demonstrates the use of a simple `DO` element, defined at the deck level.

```
<wml>
   <template>
      <do type="accept" label="Exit">
<prev/>
      </do>
   </template>
   <card>
   <p>
      Hello World!
   </p>
   </card>
</wml>
```

11.6 Control Elements

11.6.1 The Tabindex Attribute

Attributes

`tabindex=number`

This attribute specifies the tabbing position of the current element. The tabbing position indicates the relative order in which elements are traversed when tabbing within a single WML card. A numerically greater TABINDEX value indicates an element that is later in the tab sequence than an element with a numerically lesser tabindex value.

Each input element (i.e., `input` and `select`) in a card is assigned a position in the card's tab sequence. In addition, the user agent may assign a tab position to other elements. The tabindex attribute indicates the tab position of a given element. Elements that are not designated with an author-specified tab position may be assigned one by the user agent. User agent specified tab positions must be later in the tab sequence than any author-specified tab positions.

Tabbing is a navigational accelerator and is optional for all user agents. Authors must not assume that a user agent implements tabbing.

11.6.2 Select Lists

Select lists are an input element that specifies a list of options for the user to choose from. Single and multiple choice lists are supported.

11.6.2.1 The Select Element

```
<!ELEMENT select (optgroup|option)+>
<!ATTLIST select
    title       %vdata;         #IMPLIED
    name        NMTOKEN         #IMPLIED
    value       %vdata;         #IMPLIED
    iname       NMTOKEN         #IMPLIED
    ivalue      %vdata;         #IMPLIED
    multiple    %boolean;       "false"
    tabindex    %number;        #IMPLIED
    xml:lang    NMTOKEN         #IMPLIED
    %coreattrs;
>
```

The `select` element lets users pick from a list of options. Each option is specified by an `option` element. Each `option` element may have one line of formatted text (which may be wrapped or truncated by the user agent if too long). Option elements may be organised into hierarchical groups using the `optgroup` element.

Attributes
   multiple=*boolean*

This attribute indicates that the select list should accept multiple selections. When not set, the select list should only accept a single selected option.

   name=*nmtoken*
   value=*vdata*

This `name` attribute indicates the name of the variable to set with the result of the selection. The variable is set to the string value of the chosen `option` element, which is specified with the `value` attribute. The `name` variable's value is used to pre-select options in the select list.

The `value` attribute indicates the default value of the variable named in the `name` attribute. When the element is displayed, and the variable named in the `name` attribute is not set, the `name` variable may be assigned the value specified in the `value` attribute, depending on the values defined in `iname` and `ivalue`. If the `name` variable already contains a value, the `value` attribute is ignored. Any application of the default value is done before the list is pre-selected with the value of the `name` variable.

If this element allows the selection of multiple options, the result of the user's choice is a list of all selected values, separated by the semicolon character. The `name` variable is set with this result. In addition, the `value` attribute is interpreted as a semicolon-separated list of pre-selected options.

   iname=*nmtoken*
   ivalue=*vdata*

The `iname` attribute indicates the name of the variable to be set with the index result of the selection. The index result is the position of the currently selected `option` in the select list. An index of zero indicates that no `option` is selected. Index numbering begins at one and increases monotonically.

The `ivalue` attribute indicates the default-selected `option` element. When the element is displayed, if the variable named in the `iname` attribute is not set, it is assigned the default-selected entry. If the variable already contains a value, the `ivalue` attribute is ignored. If the `iname` attribute is not specified, the `ivalue` value is applied every time the element is displayed.

If this element allows the selection of multiple options, the index result of the user's choice is a list of the indices of all the selected options, separated by the semicolon character (e.g., "1;2"). The `iname` variable is set with this result. In addition, the `ivalue` attribute is interpreted as a semicolon-separated list of pre-selected options (e.g., "1;4").

    title=*vdata*

This attribute specifies a title for this element, which may be used in the presentation of this object.

Attributes defined elsewhere
    xml:lang (see section 0)
    id (see section 0)
    class (see section 0)
    tabindex (see section 0)

On entry into a card containing a `select` element, the user agent must select the initial `option` elements options in the following manner. Note that values are a semicolon delimited list of values when `multiple="true"`, but are otherwise treated as a single value (even if they contain semicolons). In addition, the default option index is an aggregate value (a list) when `multiple="true"` and is otherwise a single index.

The selection of initial `option` elements includes an operation named *validate*. This operates on a value, and determines if that value is a legal option index (or indices when `multiple="true"`). The operation consists of the following steps:

1. Remove all non-integer indices from the value.
2. Remove all out-of-range indices from the value, where out-of-range is defined as any index with a value greater than the number of options in the `select` or with a value less than one.
3. Remove duplicate indices

Note that an invalid index will result in an *empty* value.

The selection of the initial `option` elements consists of the following steps:

Step 1 - the *default option index* is determined using `iname` and `ivalue`:

- IF the `iname` attribute is specified AND names a variable that is set, THEN the default option index is the validated value of that variable.
- IF the default option index is empty AND the `ivalue` attribute is specified, THEN the default option index is the validated attribute value.
- IF the default option index is empty, AND the `name` attribute is specified AND the `name` attribute names a variable that is set, THEN for each value in the `name` variable that is present as a value in the `select`'s `option` elements, the index of the first `option` element containing that value is added to the default index if that index has not been previously added.

- IF the default option index is empty AND the `value` attribute is specified THEN for each value in the `value` attribute that is present as a value in the `select`'s `option` elements, the index of the first `option` element containing that value is added to the default index if that index has not been previously added.
- IF the default option index is empty AND the `select` is a multi-choice, THEN the default option index is set to zero.
- IF the default option index is empty AND the `select` is a single-choice, THEN the default option index is set to one.

Step 2 – initialise variables

- IF the `name` attribute is specified AND the `select` is a single-choice element, THEN the named variable is set with the value of the `value` attribute on the `option` element at the default option index.
- Else, IF the `name` attribute is specified and the `select` is a multiple-choice element, THEN for each index greater than zero, the value of the `value` attribute on the `option` element at the index is added to the `name` variable.
- IF the `iname` attribute is specified, THEN the named variable is set with the default option index.

Step 3 – pre-select option(s) specified by the default option index

- Deselect all options
- For each index greater than zero, select the option specified by the index.

When the user selects or deselects one or more `option` elements, the `name` and `iname` variables are updated with the option's value and index. The name is unset if all selected `option` elements contain an empty `value` attribute. However, in all cases, the user agent must not exhibit display side effects as a result of updating `name` and `iname` variables, except when there is an explicit refresh task (see section 9.4.3). The user agent must update `name` and `iname` variables (if specified) for each `select` element in the `card` before each and all task invocations according to steps 1 and 2 above.

Multiple choice selection lists result in a value that is a semicolon delimited list (e.g., "dog;cat"). This is not an ordered list and the user agent is free to construct the list in any order that is convenient. Authors must not rely on a particular value ordering. The user agent must ensure that the `iname` result contains no duplicate index values. The `name` result must contain duplicate values in the situation where multiple selected `option` elements have the same value. The `name` result must not contain empty values (e.g., "cat;;dog" is illegal).

CONFORMANCE RULES:

WML-44.	select	M

11.6.2.2 The Option Element

```
<!ELEMENT option (#PCDATA | onevent)*>
<!ATTLIST option
  value       %vdata;    #IMPLIED
  title       %vdata;    #IMPLIED
  onpick      %HREF;     #IMPLIED
  xml:lang    NMTOKEN    #IMPLIED
  %coreattrs;
>
```

This element specifies a single choice option in a `select` element.

Attributes

`value=vdata`

The `value` attribute specifies the value to be used when setting the `name` variable. When the user selects this option, the resulting value specified in the `value` attribute is used to set the `select` element's `name` variable.

The `value` attribute may contain variable references, which are evaluated before the `name` variable is set.

`title=vdata`

This attribute specifies a title for this element, which may be used in the presentation of this object.

`onpick=HREF`

The `onpick` event occurs when the user selects or deselects this option. A multiple-selection option list generates an `onpick` event whenever the user selects or deselects this option. A single-selection option list generates an `onpick` event when the user selects this option, i.e., no event is generated for the de-selection of any previously selected option.

Attributes defined elsewhere

`xml:lang` (see section 0)
`id` (see section 0)
`class` (see section 0)

CONFORMANCE RULES:

WML-42.	option	M

11.6.2.3 The Optgroup Element

```
<!ELEMENT optgroup (optgroup|option)+ >
<!ATTLIST optgroup
  title      %vdata;    #IMPLIED
  xml:lang   NMTOKEN    #IMPLIED
  %coreattrs;
>
```

The `optgroup` element allows the author to group related `option` elements into a hierarchy. Within a hierarchy, all leaf elements must be `option` elements (i.e., it is an error to build a hierarchy that contains a leaf `optgroup` element. The user agent may use this hierarchy to facilitate layout and presentation on a wide variety of devices. The user agent may choose not to build a hierarchy effectively ignoring `optgroup` elements. However, in all cases, the user agent must continue processes all the element's children.

Attributes

`title=vdata`

This attribute specifies a title for this element, which may be used in the presentation of this object.

Attributes defined elsewhere

`xml:lang` (see section 0)
`id` (see section 0)
`class` (see section 0)

CONFORMANCE RULES:

WML-41.   optgroup   O

11.6.2.4 Select list examples

In this example, a simple single-choice select list is specified. If the user were to choose the "Dog" option, the variable "X" would be set to a value of "D".

```
<wml>
<card>
    <p>
    Please choose your favourite animal:
        <select name="X">
            <option value="D">Dog</option>
            <option value="C">Cat</option>
        </select>
    </p>
</card>
</wml>
```

In this example, a single choice select list is specified. If the user were to choose the "Cat" option, the variable "I" would be set to a value of "2". In addition, the "Dog" option would be pre-selected if the "I" variable had not been previously set.

```
<wml>
   <card>
     <p>
     Please choose your favourite animal:
     <select iname="I" ivalue="1">
        <option value="D">Dog</option>
        <option value="C">Cat</option>
     </select>
     </p>
   </card>
</wml>
```

In this example, a multiple-choice list is specified. If the user were to choose the "Cat" and "Horse" options, the variable "X" would be set to "C;H" and the variable "I" would be set to "2;3". In addition, the "Dog" and "Cat" options would be pre-selected if the variable "I" had not been previously set.

```
<wml>
   <card>
      <p>
      Please choose <i>all</i> of your favourite animals:
      <select name="X" iname="I" ivalue="1;2" multiple="true">
         <option value="D">Dog</option>
         <option value="C">Cat</option>
         <option value="H">Horse</option>
      </select>
      </p>
   </card>
</wml>
```

In this example, a single choice select list is specified. The variable "F" would be set to the value of "S" if the user chooses the first option. The second option is always pre-selected, regardless of the value of the variable "F".

```
<wml>
   <card>
      <p>
      Please choose from the menu:
      <select name="F" ivalue="2">
         <option value="S">Sandwich</option>
         <option value="D">Drink</option>
      </select>
      </p>
   </card>
</wml>
```

In this example, the use of the `onpick` intrinsic event is demonstrated. If the user selects the second option, a `go` will be performed to the "`/morehelp.wml`" URL.

```
<wml>
   <card>
      <p>
      Select type of help:
      <select>
         <option onpick="/help.wml">Help</option>
         <option onpick="/morehelp.wml">More Help</option>
      </select>
      </p>
   </card>
</wml>
```

In this example, if the `name` variable is set to the value "`1;2`", the third option will be pre-selected. This demonstrates that values containing semicolons are treated as a single value in a single-choice selection element.

```
<wml>
<card>
   <p>
Select one:
<select name="K">
<option value="1">One</option>
<option value="2">Two</option>
<option value="1;2">Both</option>
</select>
</p>
</card>
</wml>
```

11.6.3 The Input Element

```
<!ELEMENT input EMPTY>
<!ATTLIST input
   name        NMTOKEN           #REQUIRED
   type        (text|password)   "text"
```

```
value          %vdata;         #IMPLIED
format         CDATA           #IMPLIED
emptyok        %boolean;       "false"
size           %number;        #IMPLIED
maxlength      %number;        #IMPLIED
tabindex       %number;        #IMPLIED
title          %vdata;         #IMPLIED
accesskey      %vdata;         #IMPLIED
xml:lang       NMTOKEN         #IMPLIED
%coreattrs;
>
```

The `input` element specifies a text entry object. The user input is constrained by the optional `format` attribute. If a valid input mask is bound to an input object, the user agent must ensure that any value collected by the entry object conforms to the bound input mask. If the input collected does not conform to the input mask, the user agent must not commit that input and must notify the user that the input was rejected and allow the user to resubmit new input. The user agent must not initialise the input object with any value that does not conform to the bound input mask. In the event that initialising data does not conform to the input mask, the user agent must behave as if there was no initialisation data.

Attributes

    name=nmtoken
    value=vdata

The `name` attribute specifies the name of the variable to set with the result of the user's text input. The `name` variable's value is used to pre-load the text entry object. If the `name` variable contains a value that does not conform to the input mask, the user agent must unset the variable and attempt to initialise the variable with the `value` attribute.

The `value` attribute indicates the default value of the variable named in the `name` attribute. When the element is displayed and the variable named in the `name` attribute is not set, the `name` variable is assigned the value specified in the `value` attribute. If the `name` variable already contains a value, the `value` attribute is ignored. If the `value` attribute specifies a value that does not conform to the input mask specified by the `format` attribute, the user agent must ignore the `value` attribute. In the case where no valid value can be established, the `name` variable is left unset.

    type=(text|password)

This attribute specifies the type of text-input area. The default type is `text`. The following values are allowed:

- `text` - a text entry control. User agents should echo the input in a manner appropriate to the user agent and the input mask. If the submitted value conforms to an existing input mask, the user agent must store that input unaltered and in its entirety in the variable named in the `name` attribute. For example, the user agent must not trim the input by removing leading or trailing white space from the input. If the variable named by the `name` attribute is unset, the user agent should echo an empty string in an appropriate manner.
- `password` - a text entry control. Input of each character should be echoed in an obscured or illegible form in a manner appropriate to the user agent. For example, visual user agents may elect to display an asterisk in place of a character entered by the user. Typically, the `password` input mode is indicated for password entry or other private data. Note that `Password` input is not secure and should not be depended on for critical applications. Sim-

ilar to a `text` type, if the submitted value conforms to an existing input mask, the user agent must store input unaltered and in its entirety in the variable named in the `name` attribute. User agents should not obscure non-formatting characters of the input mask. If the variable named by the `name` attribute is unset, the user agent should echo an empty string in an appropriate manner.

`format=cdata`

The FORMAT attribute specifies an input mask for user input entries. The string consists of mask control characters and static text that is displayed in the input area. The user agent may use the format mask to facilitate accelerated data input. An input mask is only valid when it contains only legal format codes. User agents must ignore invalid masks.

The format control characters specify the data format expected to be entered by the user. The default format is "*M". The format codes are:

- **A** entry of any upper-case alphabetic or punctuation character (i.e., upper-case non-numeric character).
- **a** entry of any lower-case alphabetic or punctuation character (i.e., lower-case non-numeric character).
- **N** entry of any numeric character.
- **X** entry of any upper case character.
- **x** entry of any lower-case character.
- **M** entry of any character; the user agent may choose to assume that the character is upper-case for the purposes of simple data entry, but must allow entry of any character.
- **m** entry of any character; the user agent may choose to assume that the character is lower-case for the purposes of simple data entry, but must allow entry of any character.
- ***f** entry of any number of characters; f is one of the above format codes and specifies what kind of characters can be entered. *Note: This format may only be specified once and must appear at the end of the format string.*
- **nf** entry of up to *n* characters where *n* is from 1 to 9; f is one of the above format codes (other than *f format code) and specifies what kind of characters can be entered. *Note: This format may only be specified once and must appear at the end of the format string.*
- **\c** display the next character, c, in the entry field; allows escaping of the format codes as well as introducing non-formatting characters so they can be displayed in the entry area. Escaped characters are considered part of the input's value, and must be preserved by the user agent. For example, the stored value of the input "12345-123" having a mask "NNNNN\-3N" is "12345-123" and not "12345123". Similarly, if the value of the variable named by the `name` attribute is "12345123" and the mask is "NNNNN\-3N", the user agent must unset the variable since it does not conform to the mask.

`emptyok=boolean`

The emptyok attribute indicates that this `input` element accepts empty input although a non-empty format string has been specified. Typically, the `emptyok` attribute is indicated for formatted entry fields that are optional. By default, `input` elements specifying a `format` require the user to input data matching the `format` specification.

`size=number`

This attribute specifies the width, in characters, of the text-input area. The user agent may ignore this attribute.

`maxlength=number`

This attribute specifies the maximum number of characters that can be entered by the user in the text-entry area. The default value for this attribute is an unlimited number of characters.

`title=vdata`

This attribute specifies a title for this element, which may be used in the presentation of this object.

Attributes defined elsewhere

```
xml:lang  (see section 0)
id        (see section 0)
class     (see section 0)
tabindex  (see section 0)
accesskey (see section 0)
```

CONFORMANCE RULES:

WML-33.   input   M

11.6.3.1 Input Element Examples

In this example, an `input` element is specified. This element accepts any characters and displays the input to the user in a human-readable form. The maximum number of character entered is 32 and the resulting input is assigned to the variable named X.

```
<input name="X" type="text" maxlength="32"/>
```

The following example requests input from the user and assigns the resulting input to the variable `name`. The text field has a default value of "Robert".

```
<input name="NAME" type="text" value="Robert"/>
```

The following example is a card that prompts the user for a first name, last name and age.

```
<card>
  <p>
  First name: <input type="text" name="first"/><br/>
  Last name: <input type="text" name="last"/><br/>
  Age: <input type="text" name="age" format="*N"/>
  </p>
</card>
```

11.6.4 The Fieldset Element

```
<!ELEMENT fieldset (%fields; | do)* >
<!ATTLIST fieldset
  title           %vdata;       #IMPLIED
  xml:lang        NMTOKEN       #IMPLIED
  %coreattrs;
>
```

The `fieldset` element allows the grouping of related fields and text. This grouping provides information to the user agent, allowing the optimising of layout and navigation. Fieldset elements may nest, providing the user with a means of specifying behaviour across a wide variety of devices. It is an error to include empty `fieldset` elements. See section 0 for information on how the `fieldset` element may influence layout and navigation. If a user agent chooses to discard fieldsets, it must continue to process all its children.

Attributes
```
title=vdata
```
This attribute specifies a title for this element, which may be used in the presentation of this object.

Attributes defined elsewhere
```
xml:lang  (see section 0)
id        (see section 0)
class     (see section 0)
```

CONFORMANCE RULES:

WML-28.	fieldset	O

11.6.4.1 Fieldset Element Examples

The following example specifies a WML deck that requests basic identity and personal information from the user. It is separated into multiple field sets, indicating the preferred field grouping to the user agent.

```
<wml>
<card>
    <p>
    <do type="accept">
       <go href="/submit?f=$(fname)&amp;l=$(lname)&amp;s=$(sex)&amp;a=$(age)"/>
    </do>
    <fieldset title="Name">
        First name: <input type="text" name="fname" maxlength="32"/>
<br/>Last name: <input type="text" name="lname" maxlength="32"/>
    </fieldset>
    <fieldset title="Info">
        <select name="sex">
           <option value="F">Female</option>
           <option value="M">Male</option>
        </select>
        <br/>
        Age: <input type="text" name="age" format="*N"/>
    </fieldset>
    </p>
  </card>
</wml>
```

11.7 The Timer Element

```
<!ELEMENT timer EMPTY>
<!ATTLIST timer
   name     NMTOKEN       #IMPLIED
   value    %vdata;       #REQUIRED
```

```
    %coreattrs;
>
```

The `timer` element declares a card timer, which exposes a means of processing inactivity or idle time. The timer is initialised and started at card entry and is stopped when the card is exited. Card entry is any task or user action that results in the card being activated, for example, navigating into the card. Card exit is defined as the execution of any task (see sections 0 and 0). The value of a timer will decrement from the initial value, triggering the delivery of an `ontimer` intrinsic event on transition from a value of one to zero. If the user has not exited the card at the time of timer expiration, an `ontimer` intrinsic event is delivered to the card.

Timer resolution is implementation dependent. The interaction of the timer with the user agent's user interface and other time-based or asynchronous device functionality is implementation dependent. It is an error to have more than one `timer` element in a card.

The `timer` timeout value is specified in units of one-tenth (1/10) of a second. The author should not expect a particular timer resolution and should provide the user with another means to invoke a timer's task. If the value of the timeout is not a positive integral number, the user agent must ignore the `timer` element. A timeout value of zero (0) disables the timer.

Invoking a refresh task is considered an exit. The task stops the timer, commits it's value to the context, and updates the user agent accordingly. Completion of the refresh task is considered an entry to the card. At that time, the timer must resume.

Attributes

`name=`*nmtoken*

The `name` attribute specifies the name of the variable to be set with the value of the timer. The `name` variable's value is used to set the timeout period upon timer initialisation. The variable named by the `name` attribute will be set with the current timer value when the card is exited or when the timer expires. For example, if the timer expires, the `name` variable is set to a value of "0".

`value=`*vdata*

The `value` attribute indicates the default value of the variable named in the `name` attribute. When the timer is initialised and the variable named in the `name` attribute is not set, the `name` variable is assigned the value specified in the `value` attribute. If the `name` variable already contains a value, the `value` attribute is ignored. If the `name` attribute is not specified, the timeout is always initialised to the value specified in the `value` attribute.

Attributes defined elsewhere

`id` (see section 0)
`class` (see section 0)

CONFORMANCE RULES:

WML-49.   timer   M

11.7.1 Timer Example

The following deck will display a text message for approximately 10 seconds and will then go to the URL `/next`.

```
<wml>
<card ontimer="/next">
    <timer value="100"/>
```

```
            <p>
            Hello World!
            </p>
</card>
</wml>
```

The same example could be implemented as:

```
<wml>
<card>
<onevent type="ontimer">
<go href="/next"/>
        </onevent>
<timer value="100"/>
<p>
Hello World!
</p>
</card>
</wml>
```

The following example illustrates how a timer can initialise and reuse a counter. Each time the card is entered, the timer is reset to value of the variable `t`. If `t` is not set, the timer is set to a value of 5 seconds.

```
<wml>
<card ontimer="/next">
<timer name="t" value="50"/>
<p>
Hello World!
</p>
</card>
</wml>
```

11.8 Text

This section defines the elements and constructs related to text.

11.8.1 White Space

WML white space and line break handling is based on [XML] and assumes the default XML white space handling rules for text. The WML user agent ignores all *insignificant* white space in elements and attribute values, as defined by the XML specification. White space immediately before and after an element is ignored. In addition, all other sequences of white space must be compressed into a single inter-word space.

User agents should treat inter-word spaces in a locale-dependent manner, as different written languages treat inter-word spacing in different ways.

11.8.2 Emphasis

```
<!ELEMENT em       (%flow;)*>
<!ATTLIST em
  xml:lang         NMTOKEN         #IMPLIED
  %coreattrs;
```

```
  >
<!ELEMENT strong   (%flow;)*>
<!ATTLIST strong
  xml:lang         NMTOKEN       #IMPLIED
  %coreattrs;
  >
<!ELEMENT i        (%flow;)*>
<!ATTLIST i
  xml:lang         NMTOKEN       #IMPLIED
  %coreattrs;
  >
<!ELEMENT b        (%flow;)*>
<!ATTLIST b
  xml:lang         NMTOKEN       #IMPLIED
  %coreattrs;
  >
<!ELEMENT u        (%flow;)*>
<!ATTLIST u
  xml:lang         NMTOKEN       #IMPLIED
  %coreattrs;
  >
<!ELEMENT big      (%flow;)*>
<!ATTLIST big
  xml:lang         NMTOKEN       #IMPLIED
  %coreattrs;
  >
<!ELEMENT small    (%flow;)*>
<!ATTLIST small
  xml:lang         NMTOKEN       #IMPLIED
  %coreattrs;
  >
```

The emphasis elements specify text emphasis markup information.

em:
Render with emphasis.

strong:
Render with strong emphasis.

i:
Render with an italic font.

b:
Render with a bold font.

u:
Render with underline.

big:
Render with a large font.

small:
Render with a small font.

Authors should use the `strong` and `em` elements where possible. The `b`, `i` and `u` elements should not be used except where explicit control over text presentation is required.

Visual user agents must distinguish emphasised text from non-emphasised text. A user agent should do a best effort to distinguish the various forms of emphasised text as described above. It should distinguish text that has been emphasised using the `em` element from that using `strong` element. User agents may use the same style for `strong`, `b`, and `big` emphasis. It may also use the same style for `em`, `i`, `u`, and `small` emphasis.

Attributes defined elsewhere

   `xml:lang` (see section 0)
   `id` (see section 0)
   `class` (see section 0)

CONFORMANCE RULES:

WML-22.	b	O
WML-23.	big	O
WML-27.	em	O
WML-31.	i	O
WML-45.	small	O
WML-46.	strong	O
WML-52.	u	O

11.8.3 Paragraphs

```
<!ENTITY % TAlign "(left|right|center)">
<!ENTITY % WrapMode "(wrap|nowrap)" >
<!ELEMENT p (%fields; | do)*>
<!ATTLIST p
  align         %TAlign;      "left"
  mode          %WrapMode;    #IMPLIED
  xml:lang      NMTOKEN       #IMPLIED
  %coreattrs;
>
```

WML has two line-wrapping modes for visual user agents: breaking (or wrapping) and non-breaking (or non-wrapping). The treatment of a line too long to fit on the screen is specified by the current line-wrap mode. If `mode="wrap"` is specified, the line is word-wrapped onto multiple lines. In this case, line breaks should be inserted into a text flow as appropriate for presentation on an individual device. If `mode="nowrap"` is specified, the line is not automatically wrapped. In this case, the user agent must provide a mechanism to view entire non-wrapped lines (e.g., horizontal scrolling or some other user-agent-specific mechanism).

Any inter-word space is a legal line break point. The non-breaking space entity (` ` or ` `) indicates a space that must not be treated as an inter-word space by the user agent. Authors should use ` ` to prevent undesired line-breaks. The soft-hyphen character entity (`&shy;` or `&#173;`) indicates a location that may be used by the user agent for a line break. If a line break occurs at a soft-hyphen, the user agent must insert a hyphen character (`&#45;`) at the end of the line. In all other operations, the soft-hyphen entity should be ignored. A user agent may choose to entirely ignore soft-hyphens when formatting text lines.

The p element establishes both the line wrap and alignment parameters for a paragraph. If the text alignment is not specified, it defaults to left. If the line-wrap mode is not specified, it is identical to the line-wrap mode of the previous paragraph in the current card. Empty paragraphs (i.e., an empty element or an element with only insignificant white space) should be considered as insignificant and ignored by visual user agents. Insignificant paragraphs do not impact line-wrap mode. If the first p element in a card does not specify a line-wrap or alignment mode, that mode defaults to the initial mode for the card. The user agent must insert a line break into the text flow between significant p elements.

Insignificant paragraphs may be removed before the document is delivered to the user agent.

Attributes

```
align=(left|right|center)
```

This attribute specifies the text alignment mode for the paragraph. Text can be centre aligned, left aligned or right aligned when it is displayed to the user. Left alignment is the default alignment mode. If not explicitly specified, the text alignment is set to the default alignment.

```
mode=(wrap|nowrap)
```

This attribute specifies the line-wrap mode for the paragraph. Wrap specifies breaking text mode and nowrap specifies non-breaking text mode. If not explicitly specified, the line-wrap mode is identical to the line-wrap mode of the previous paragraph in the text flow of a card. The default mode for the first paragraph in a card is wrap.

Attributes defined elsewhere

xml:lang (see section 0)
id (see section 0)
class (see section 0)

CONFORMANCE RULES:

WML-36.   p   M

11.8.4 The Br Element

```
<!ELEMENT br EMPTY>
<!ATTLIST br
  %coreattrs;
  >
```

The br element establishes the beginning of a new line. The user agent must break the current line and continue on the following line. User agents should do best effort to support the br element in tables (see section 0).

Attributes defined elsewhere

id (see section 0)
class (see section 0)

CONFORMANCE RULES:

WML-24.   br   M

11.8.5 The Table Element

```
<!ELEMENT table (tr)+>
<!ATTLIST table
  title      %vdata;     #IMPLIED
```

```
           align        CDATA        #IMPLIED
           columns      %number;     #REQUIRED
           xml:lang     NMTOKEN      #IMPLIED
           %coreattrs;
   >
```

The `table` element is used together with the `tr` and `td` elements to create sets of aligned columns of text and images in a card. Nesting of `table` elements is not allowed. The `table` elements determine the structure of the columns. The elements separate content into columns, but do not specify column or intercolumn widths. The user agent should do its best effort to present the information of the table in a manner appropriate to the device.

Attributes

`title=`*vdata*

This attribute specifies a title for this element, which may be used in the presentation of this object.

`align=`*cdata*

This attribute specifies the layout of text and images within the columns of a table. A column's contents can be centre aligned, left aligned or right aligned when it is rendered to the user. The attribute value is interpreted as a non-separated list of alignment designations, one for each column. Centre alignment is specified with the value "C", left alignment is specified with the value "L", right alignment is specified with the value "R", and default alignment is specified with the value "D". Designators are applied to columns as they are defined in the content. The first designator in the list applies to the first column, the second designator to the second column, and so forth. Default alignment is applied to columns that are missing alignment designators or have unrecognised designators. All extra designators are ignored. Determining the default alignment is implementation dependent. User agents should consider the current language when determining the default alignment and the direction of the table. A user agent may use other algorithms to make such decisions.

`columns=`*number*

This required attribute specifies the number of columns for the table. The user agent must create a table with exactly the number of columns specified by the attribute value. It is an error to specify a value of zero ("0").

If the actual number of columns in a row is less than the value specified by the columns attribute, the row must be padded with empty columns effectively as if the user agent appended empty `td` elements to the row.

If the actual number of columns in a row is greater than the value specified by this attribute, the extra columns of the row must be aggregated into the last column such that the row contains exactly the number of columns specified. A single inter-word space must be inserted between two cells that are being aggregated.

The presentation of the table is likely to depend on the display characteristics of the device. WML does not define how a user agent renders a table. User agents may create aligned columns for each table, or it may use a single set of aligned columns for all tables in a card. User agents that choose to render a table in a traditional tabular manner should determine the width of each column from the maximum width of the text and images in that column to ensure the narrowest display width. However, user agents may use fixed width or other appropriate layout algorithms instead. User agents that choose to render tables in a traditional tabular manner must use a non-zero width gutter to separate each non-empty column.

Attributes defined elsewhere
    xml:lang (see section 0)
    id (see section 0)
    class (see section 0)

CONFORMANCE RULES:

WML-47.	table	M
WML-63.	The number of columns in a table must not be set to zero	M

11.8.6 The Tr Element

```
<!ELEMENT tr (td)+>
<!ATTLIST tr
  %coreattrs;
  >
```

The `tr` element is used as a container to hold a single table row. Table rows may be empty (i.e., all cells are empty). Empty table rows are significant and must not be ignored.

Attributes defined elsewhere
    id (see section 0)
    class (see section 0)

CONFORMANCE RULES:

WML-51.	tr	M

11.8.7 The Td Element

```
<!ELEMENT td ( %text; | %layout; | img | anchor | a )*>
<!ATTLIST td
  xml:lang      NMTOKEN      #IMPLIED
  %coreattrs;
  >
```

The `td` element is used as a container to hold a single table cell data within a table row. Table data cells may be empty. Empty cells are significant, and must not be ignored. The user agent should do a best effort to deal with multiple line data cells that may result from using images or line breaks.

Attributes defined elsewhere
    xml:lang (see section 0)
    id (see section 0)
    class (see section 0)

CONFORMANCE RULES:

WML-50.	td	M

11.8.8 Table Example

The following example contains a card with a single column group, containing two columns and three rows.

```
<wml>
  <card>
    <p>
```

```
            <table columns="2" align="LL">
                <tr><td>One </td><td> Two </td></tr>
                <!-- row missing cells -->
<tr><td>1</td></tr>
                <!-- row with too many cells -->
<tr><td/><td> B </td><td>C<br/>D</td></tr>
            </table>
        </p>
    </card>
</wml>
```

An acceptable layout for this card is:

```
One     Two
1
        B C
        D
```

11.8.9 The Pre Element

```
<!ELEMENT pre "(#PCDATA | a | br | i | b | em | strong | input | select )*">
<!ATTLIST pre
  xml:space    CDATA    #FIXED "preserve"
  %coreattrs;
  >
```

The PRE element tells visual user agents that the enclosed text is "preformatted". When handling preformatted text, user agents:

May leave white space intact.

May render text with a fixed-pitch font.

May disable automatic word wrap.

The user agent must make "best effort" to achieve the requirements above.

CONFORMANCE RULES:

WML-38.   pre   O

11.9 Images

```
<!ENTITY % IAlign "(top|middle|bottom)" >
<!ELEMENT img EMPTY>
<!ATTLIST img
  alt          %vdata;        #REQUIRED
  src          %HREF;         #REQUIRED
  localsrc     %vdata;        #IMPLIED
  vspace       %length;       "0"
  hspace       %length;       "0"
  align        %IAlign;       "bottom"
  height       %length;       #IMPLIED
  width        %length;       #IMPLIED
  xml:lang     NMTOKEN        #IMPLIED
  %coreattrs;
  >
```

The `img` element indicates that an image is to be included in the text flow. Image layout is done within the context of normal text layout.

Attributes

`alt=vdata`

This attribute specifies an alternative textual representation for the image. This representation is used when the image can not be displayed using any other method (i.e., the user agent does not support images, or the image contents can not be found).

`src=HREF`

This attribute specifies the URI for the image. If the browser supports images, it downloads the image from the specified URI and renders it when the text is being displayed.

`localsrc=vdata`

This attribute specifies an alternative internal representation for the image. This representation is used if it exists; otherwise the image is downloaded from the URI specified in the `src` attribute, i.e., any `localsrc` parameter specified takes precedence over the image specified in the `src` parameter.

`vspace=length`
`hspace=length`

These attributes specify the amount of white space to be inserted to the left and right (`hspace`) and above and below (`vspace`) the image. The default value for this attribute is zero indicating that no white space should be inserted. If `length` is specified as a percentage value, the space inserted is based on the available horizontal or vertical space. These attributes are hints to the user agent and may be ignored.

`align=(top|middle|bottom)`

This attribute specifies image alignment within the text flow and with respect to the current insertion point. `align` has three possible values:

- `bottom` - means that the bottom of the image should be vertically aligned with the current baseline. This is the default value.
- `middle` - means that the centre of the image should be vertically aligned with the centre of the current text line.
- `top` - means that the top of the image should be vertically aligned with the top of the current text line.

`height=length`
`width=length`

These attributes give user agents an idea of the size of an image or object so that they may reserve space for it and continue rendering the card while waiting for the image data. User agents may scale objects and images to match these values if appropriate. If `length` is specified as a percentage value, the resulting size is based on the available horizontal or vertical space, not on the natural size of the image. These attributes are a hint to the user agent and may be ignored.

Attributes defined elsewhere

`xml:lang` (see section 0)
`id` (see section 0)
`class` (see section 0)

CONFORMANCE RULES:

WML-32.	img	M

12. User Agent Semantics

Except where explicitly stated WML does not dictate how a user agent should render or display WML content. The user agent is not obligated to perform any particular mapping of elements to user interface widgets, and the WML author should not rely on such.

12.1 Deck Access Control

The introduction of variables into WML exposes potential security issues that do not exist in other markup languages such as HTML. In particular, certain variable state may be considered private by the user. While the user may be willing to send a private information to a secure service, an insecure or malicious service should not be able to retrieve that information from the user agent by other means.

A conforming WML user agent must implement deck-level access control, including the `access` element and the `sendreferer`, `domain` and `path` attributes.

A WML author should remove private or sensitive information from the browser context by clearing the variables containing this information.

CONFORMANCE RULES:

WML-14.	Deck access control	M

12.2 Low-Memory Behaviour

WML is targeted at devices with limited hardware resources, including significant restrictions on memory size. It is important that the author have a clear expectation of device behaviour in error situations, including those caused by lack of memory.

CONFORMANCE RULES:

WML-15.	Low-memory	O

12.2.1 Limited History

The user agent may limit the size of the history stack (i.e., the depth of the historical navigation information). In the case of history size exhaustion, the user agent should delete the least-recently-used history information.

It is recommended that all user agents implement a minimum history stack size of ten entries.

12.2.2 Limited Browser Context Size

In some situations, it is possible that the author has defined an excessive number of variables in the browser context, leading to memory exhaustion.

In this situation, the user agent should attempt to acquire additional memory by reclaiming cache and history memory as described in sections 0. If this fails and the user agent has exhausted all memory, the user should be notified of the error, and the user agent should be reset to a predictable user state. For example, the browser may be terminated or the context may be cleared and the browser reset to a well-known state.

© Copyright Wireless Application Protocol Forum, Ltd., 1998, 1999
All rights reserved

12.3 Error Handling

Conforming user agents must enforce error conditions defined in this specification and must not hide errors by attempting to infer author or origin server intent.

CONFORMANCE RULES:

WML-16.   Error handling   M

12.4 Unknown DTD

A WML deck encoded with an alternate DTD may include elements or attributes that are not recognised by certain user agents. In this situation, a user agent should render the deck as if the unrecognised tags and attributes were not present. Content contained in unrecognised elements should be rendered.

CONFORMANCE RULES:

WML-17.   Unknown DTD handling   M

12.5 Reference Processing Behaviour - Inter-card Navigation

The following process describes the reference model for inter-card traversal in WML. All user agents must implement this process, or one that is indistinguishable from it.

CONFORMANCE RULES:

WML-18.   Inter-card navigation   M

12.5.1 The Go Task

The process of executing a `go` task comprises the following steps:

1. If the originating task contains `setvar` elements, the variable name and value in each `setvar` element is converted into a simple string by substituting all referenced variables. The resulting collection of variable names and values is stored in temporary memory for later processing. See section 0 for more information on variable substitution.
2. The target URI is identified and fetched by the user agent. The URI attribute value is converted into a simple string by substituting all referenced variables.
3. The access control parameters for the fetched deck are processed as specified in section 0.
4. The destination card is located using the fragment name specified in the URI.
    a. If no fragment name was specified as part of the URI, the first card in the deck is the destination card.
    b. If a fragment name was identified and a card has a `name` attribute that is identical to the fragment name, then that card is the destination card.
    c. If the fragment name can not be associated with a specific card, the first card in the deck is the destination card.
5. The variable assignments resulting from the processing done in step #1 (the `setvar` element) are applied to the current browser context.
6. If the destination card contains a `newcontext` attribute, the current browser context is re-initialised as described in section 0.
7. The destination card is pushed onto the history stack.

8. If the destination card specifies an `onenterforward` intrinsic event binding, the task associated with the event binding is executed and processing stops. See section 0 for more information.
9. If the destination card contains a `timer` element, the timer is started as specified in section 0.

The destination card is displayed using the current variable state and processing stops.

12.5.2 The Prev Task

The process of executing a `prev` task comprises the following steps:
1. If the originating task contains `setvar` elements, the variable name and value in each `setvar` element is converted into a simple string by substituting all referenced variables. The resulting collection of variable names and values is stored in temporary memory for later processing. See section 0 for more information on variable substitution.
2. The target URI is identified and fetched by the user agent. The history stack is popped and the target URI is the top of the history stack. If there is no previous card in the history stack, processing stops.
3. The destination card is located using the fragment name specified in the URI.
   a. If no fragment name was specified as part of the URI, the first card in the deck is the destination card.
   b. If a fragment name was identified and a card has a `name` attribute that is identical to the fragment name, then that card is the destination card.
4. The variable assignments resulting from the processing done in step #1 (the `setvar` element) are applied to the current browser context.
5. If the destination card specifies an `onenterbackward` intrinsic event binding, the task associated with the event binding is executed and processing stops. See section 0 for more information.
6. If the destination card contains a `timer` element, the timer is started as specified in section 0.
7. The destination card is displayed using the current variable state and processing stops.

12.5.3 The Noop Task

No processing is done for a `noop` task.

12.5.4 The Refresh Task

The process of executing a `refresh` task comprises the following steps:
1. For each `setvar` element, the variable name and value in each `setvar` element is converted into a simple string by substituting all referenced variables. See section 0 for more information on variable substitution.
2. The variable assignments resulting from the processing done in step #1 (the `setvar` element) are applied to the current browser context.
3. If the card contains a `timer` element, the timer is started as specified in section 0.
4. The current card is re-displayed using the current variable state and processing stops.

12.5.5 Task Execution Failure

If a task fails to fetch its target URI or the access control restrictions prevent a successful inter-card transition, the user agent must notify the user and take the following actions:

- The invoking card remains the current card.
- No changes are made to the browser context, including any pending variable assignments or `newcontext` processing.
- No intrinsic event bindings are executed.

13. WML Reference Information

WML is an application of [XML] version 1.0.

13.1 Document Identifiers

13.1.1 SGML Public Identifier

```
-//WAPFORUM//DTD WML 1.2//EN
```

13.1.2 WML Media Type

Textual form:

```
text/vnd.wap.wml
```

Tokenised form:

```
application/vnd.wap.wmlc
```

13.2 Document Type Definition (DTD)

```
<!--
Wireless Markup Language (WML) Document Type Definition.
WML is an XML language.   Typical usage:
   <?xml version="1.0"?>
   <!DOCTYPE wml PUBLIC "-//WAPFORUM//DTD WML 1.2//EN"
         "http://www.wapforum.org/DTD/wml12.dtd">
   <wml>
   ...
   </wml>
-->

<!ENTITY % length    "CDATA">    <!-- [0-9]+ for pixels or [0-9]+"%" for
                                      percentage length -->
<!ENTITY % vdata     "CDATA">    <!-- attribute value possibly containing
                                      variable references -->
<!ENTITY % HREF      "%vdata;">  <!-- URI, URL or URN designating a hypertext
                                      node. May contain variable references -->
<!ENTITY % boolean   "(true|false)">
<!ENTITY % number    "NMTOKEN">  <!-- a number, with format [0-9]+ -->
<!ENTITY % coreattrs "id     ID       #IMPLIED
                      class  CDATA    #IMPLIED">
```

```
<!ENTITY % ContentType "%vdata;">  <!-- media type. May contain variable
                                        references -->

<!ENTITY % emph      "em | strong | b | i | u | big | small">
<!ENTITY % layout    "br">

<!ENTITY % text      "#PCDATA | %emph;">

<!-- flow covers "card-level" elements, such as text and images -->
<!ENTITY % flow      "%text; | %layout; | img | anchor | a | table">

<!-- Task types -->
<!ENTITY % task     "go | prev | noop | refresh">

<!-- Navigation and event elements -->
<!ENTITY % navelmts "do | onevent">


<!--================= Decks and Cards =================-->

<!ELEMENT wml ( head?, template?, card+ )>
<!ATTLIST wml
   xml:lang         NMTOKEN         #IMPLIED
   %coreattrs;
   >

<!-- card intrinsic events -->
<!ENTITY % cardev
 "onenterforward  %HREF;          #IMPLIED
  onenterbackward %HREF;          #IMPLIED
  ontimer         %HREF;          #IMPLIED"
  >

<!-- card field types -->
<!ENTITY % fields    "%flow; | input | select | fieldset">

<!ELEMENT card (onevent*, timer?, (do | p | pre)*)>
<!ATTLIST card
   title            %vdata;         #IMPLIED
   newcontext       %boolean;       "false"
   ordered          %boolean;       "true"
   xml:lang         NMTOKEN         #IMPLIED
   %cardev;
   %coreattrs;
   >

<!--================= Event Bindings =================-->

<!ELEMENT do (%task;)>
<!ATTLIST do
   type             CDATA           #REQUIRED
   label            %vdata;         #IMPLIED
   name             NMTOKEN         #IMPLIED
   optional         %boolean;       "false"
```

```
    xml:lang        NMTOKEN         #IMPLIED
    %coreattrs;
    >

<!ELEMENT onevent (%task;)>
<!ATTLIST onevent
    type            CDATA           #REQUIRED
    %coreattrs;
    >

<!--================ Deck-level declarations ================-->

<!ELEMENT head ( access | meta )+>
<!ATTLIST head
    %coreattrs;
    >

<!ELEMENT template (%navelmts;)*>
<!ATTLIST template
    %cardev;
    %coreattrs;
    >

<!ELEMENT access EMPTY>
<!ATTLIST access
    domain          CDATA           #IMPLIED
    path            CDATA           #IMPLIED
    %coreattrs;
    >

<!ELEMENT meta EMPTY>
<!ATTLIST meta
    http-equiv      CDATA           #IMPLIED
    name            CDATA           #IMPLIED
    forua           %boolean;       "false"
    content         CDATA           #REQUIRED
    scheme          CDATA           #IMPLIED
    %coreattrs;
    >

<!--================ Tasks ================-->

<!ELEMENT go (postfield | setvar)*>
<!ATTLIST go
    href            %HREF;          #REQUIRED
    sendreferer     %boolean;       "false"
    method          (post|get)      "get"
    enctype         %ContentType;   "application/x-www-form-urlencoded"
    accept-charset  CDATA           #IMPLIED
    %coreattrs;
    >

<!ELEMENT prev (setvar)*>
```

```
<!ATTLIST prev
  %coreattrs;
  >

<!ELEMENT refresh (setvar)*>
<!ATTLIST refresh
  %coreattrs;
  >

<!ELEMENT noop EMPTY>
<!ATTLIST noop
  %coreattrs;
  >

<!--=============== postfield ================-->

<!ELEMENT postfield EMPTY>
<!ATTLIST postfield
  name         %vdata;          #REQUIRED
  value        %vdata;          #REQUIRED
  %coreattrs;
  >

<!--=============== variables ================-->

<!ELEMENT setvar EMPTY>
<!ATTLIST setvar
  name         %vdata;          #REQUIRED
  value        %vdata;          #REQUIRED
  %coreattrs;
  >

<!--=============== Card Fields ================-->

<!ELEMENT select (optgroup|option)+>
<!ATTLIST select
  title        %vdata;          #IMPLIED
  name         NMTOKEN          #IMPLIED
  value        %vdata;          #IMPLIED
  iname        NMTOKEN          #IMPLIED
  ivalue       %vdata;          #IMPLIED
  multiple     %boolean;        "false"
  tabindex     %number;         #IMPLIED
  xml:lang     NMTOKEN          #IMPLIED
  %coreattrs;
  >

<!ELEMENT optgroup (optgroup|option)+ >
<!ATTLIST optgroup
  title        %vdata;    #IMPLIED
  xml:lang     NMTOKEN    #IMPLIED
  %coreattrs;
  >
```

```
<!ELEMENT option (#PCDATA | onevent)*>
<!ATTLIST option
  value       %vdata;         #IMPLIED
  title       %vdata;         #IMPLIED
  onpick      %HREF;          #IMPLIED
  xml:lang    NMTOKEN         #IMPLIED
  %coreattrs;
  >

<!ELEMENT input EMPTY>
<!ATTLIST input
  name        NMTOKEN             #REQUIRED
  type        (text|password)     "text"
  value       %vdata;             #IMPLIED
  format      CDATA               #IMPLIED
  emptyok     %boolean;           "false"
  size        %number;            #IMPLIED
  maxlength   %number;            #IMPLIED
  tabindex    %number;            #IMPLIED
  title       %vdata;             #IMPLIED
  accesskey   %vdata;             #IMPLIED
  xml:lang    NMTOKEN             #IMPLIED
  %coreattrs;
  >

<!ELEMENT fieldset (%fields; | do)* >
<!ATTLIST fieldset
  title       %vdata;         #IMPLIED
  xml:lang    NMTOKEN         #IMPLIED
  %coreattrs;
  >

<!ELEMENT timer EMPTY>
<!ATTLIST timer
  name        NMTOKEN         #IMPLIED
  value       %vdata;         #REQUIRED
  %coreattrs;
  >

<!--================ Images ================-->

<!ENTITY % IAlign "(top|middle|bottom)" >

<!ELEMENT img EMPTY>
<!ATTLIST img
  alt         %vdata;         #REQUIRED
  src         %HREF;          #REQUIRED
  localsrc    %vdata;         #IMPLIED
  vspace      %length;        "0"
  hspace      %length;        "0"
  align       %IAlign;        "bottom"
  height      %length;        #IMPLIED
```

```
    width       %length;       #IMPLIED
    xml:lang    NMTOKEN        #IMPLIED
    %coreattrs;
    >

<!--================ Anchor ================-->

<!ELEMENT anchor ( #PCDATA | br | img | go | prev | refresh )*>
<!ATTLIST anchor
    title       %vdata;        #IMPLIED
    accesskey   %vdata;        #IMPLIED
    xml:lang    NMTOKEN        #IMPLIED
    %coreattrs;
    >

<!ELEMENT a ( #PCDATA | br | img )*>
<!ATTLIST a
    href        %HREF;         #REQUIRED
    title       %vdata;        #IMPLIED
    accesskey   %vdata;        #IMPLIED
    xml:lang    NMTOKEN        #IMPLIED
    %coreattrs;
    >

<!--================ Tables ================-->

<!ELEMENT table (tr)+>
<!ATTLIST table
    title       %vdata;        #IMPLIED
    align       CDATA          #IMPLIED
    columns     %number;       #REQUIRED
    xml:lang    NMTOKEN        #IMPLIED
    %coreattrs;
    >

<!ELEMENT tr (td)+>
<!ATTLIST tr
    %coreattrs;
    >

<!ELEMENT td ( %text; | %layout; | img | anchor | a )*>
<!ATTLIST td
    xml:lang         NMTOKEN        #IMPLIED
    %coreattrs;
    >

<!--================ Text layout and line breaks ================-->

<!ELEMENT em       (%flow;)*>
<!ATTLIST em
    xml:lang         NMTOKEN        #IMPLIED
    %coreattrs;
    >
```

```
<!ELEMENT strong  (%flow;)*>
<!ATTLIST strong
  xml:lang        NMTOKEN         #IMPLIED
  %coreattrs;
  >

<!ELEMENT b       (%flow;)*>
<!ATTLIST b
  xml:lang        NMTOKEN         #IMPLIED
  %coreattrs;
  >

<!ELEMENT i       (%flow;)*>
<!ATTLIST i
  xml:lang        NMTOKEN         #IMPLIED
  %coreattrs;
  >

<!ELEMENT u       (%flow;)*>
<!ATTLIST u
  xml:lang        NMTOKEN         #IMPLIED
  %coreattrs;
  >

<!ELEMENT big     (%flow;)*>
<!ATTLIST big
  xml:lang        NMTOKEN         #IMPLIED
  %coreattrs;
  >

<!ELEMENT small   (%flow;)*>
<!ATTLIST small
  xml:lang        NMTOKEN         #IMPLIED
  %coreattrs;
  >

<!ENTITY % TAlign "(left|right|center)">
<!ENTITY % WrapMode "(wrap|nowrap)" >
<!ELEMENT p (%fields; | do)*>
<!ATTLIST p
  align         %TAlign;        "left"
  mode          %WrapMode;      #IMPLIED
  xml:lang      NMTOKEN         #IMPLIED
  %coreattrs;
  >

<!ELEMENT br EMPTY>
<!ATTLIST br
  %coreattrs;
  >

<!ELEMENT pre "(#PCDATA | a | br | i | b | em | strong | input | select )*">
```

```
<!ATTLIST pre
  xml:space    CDATA    #FIXED "preserve"
  %coreattrs;
>

<!ENTITY quot  "&#34;">       <!-- quotation mark -->
<!ENTITY amp   "&#38;#38;">   <!-- ampersand -->
<!ENTITY apos  "&#39;">       <!-- apostrophe -->
<!ENTITY lt    "&#38;#60;">   <!-- less than -->
<!ENTITY gt    "&#62;">       <!-- greater than -->
<!ENTITY nbsp  " ">      <!-- non-breaking space -->
<!ENTITY shy   "&#173;">      <!-- soft hyphen (discretionary hyphen) -->
```

13.3 Reserved Words

WML reserves the use of several strings for future uses. These strings may not be used in any DTD or extension of WML. The following words are reserved:

```
style
```

14. A Compact Binary Representation of WML

WML may be encoded using a compact binary representation. This content format is based upon the WAP Binary XML Content Format [WBXML].

CONFORMANCE RULES:

WML-55.   WML token table   M

14.1 Extension Tokens

14.1.1 Global Extension Tokens

The [WBXML] global extension tokens are used to represent WML variables. Variable references may occur in a variety of places in a WML deck (see section 0). There are several codes that indicate variable substitution. Each code has different escaping semantics (e.g., direct substitution, escaped substitution and unescaped substitution). The variable name is encoded in the current document character encoding and must be encoded as the specified in the source document (e.g., variable names may not be shortened, mapped or otherwise changed). For example, the global extension token EXT_I_0 represents an escaped variable substitution, with the variable name inline.

14.1.2 Tag Tokens

WML defines a set of single-byte tokens corresponding to the tags defined in the DTD. All of these tokens are defined within code page zero.

14.1.3 Attribute Tokens

WML defines a set of single-byte tokens corresponding to the attribute names and values defined in the DTD. All of these tokens are defined within code page zero.

14.2 Encoding Semantics

14.2.1 Encoding Variables

All valid variable references must be converted to variable reference tokens (e.g., `EXT_I_0`). The encoder must validate that a variable reference uses proper syntax. The encoder should also validate that the placement of the variable reference within the WML deck is valid.

14.2.2 Encoding Tag and Attributes Names

All tag and attribute names, for which binary token values are defined in this specification, must be tokenised, literal tokens must not be used. The user-agent must, however, treat literal and binary tokens as equivalent. See [WBXML].

CONFORMANCE RULES:

WML-56.	XML Well-formed	M
WML-57.	XML Validation	O

14.2.3 Document Validation

XML document validation (see [XML]) should occur during the process of tokenising a WML deck and must be based on the DOCTYPE declared in the WML deck. When validating the source text, the tokenisation process must accept any DOCTYPE or public identifier, if the document is identified as a WML media type (see section 0).

The tokenisation process should notify the user of any well-formedness or validity errors detected in the source deck.

14.2.3.1 Validate %length;

The WML tokenisation process should validate that attribute values defined as `%length;` contain either a `NMTOKEN` or a `NMTOKEN` followed by a percentage sign character. For example, the following attributes are legal:

```
vspace="100%"
hspace="123"
```

`%length;` data is encoded using normal attribute value encoding methods.

14.2.3.2 Validate %vdata;

The WML tokenisation process must validate the syntax of all variable references within attribute values defined as `%vdata;` or `%HREF;` according to section 0. It must also verify that other `CDATA` attribute values do not contain any variable references. Attribute values not defined in the DTD must be treated as `%vdata;` and validated accordingly.

14.3 Numeric Constants

14.3.1 WML Extension Token Assignment

The following global extension tokens are used in WML and occupy document-type-specific token slots in the global token range. As with all tokens in the global range, these codes must be reserved in every code page. All numbers are in hexadecimal.

Table 4. Global Extension Token Assignments

TOKEN NAME	TOKEN	DESCRIPTION
EXT_I_0	40	Variable substitution - escaped. Name of the variable is inline and follows the token as a `termstr`.
EXT_I_1	41	Variable substitution - unescaped. Name of the variable is inline and follows the token as a `termstr`.
EXT_I_2	42	Variable substitution - no transformation. Name of the variable is inline and follows the token as a `termstr`.
EXT_T_0	80	Variable substitution - escaped. Variable name encoded as a reference into the string table.
EXT_T_1	81	Variable substitution - unescaped. Variable name encoded as a reference into the string table.
EXT_T_2	82	Variable substitution - no transformation. Variable name encoded as a reference into the string table.
EXT_0	C0	Reserved for future use.
EXT_1	C1	Reserved for future use.
EXT_2	C2	Reserved for future use.

14.3.2 Tag Tokens

The following token codes represent tags in code page zero (0). All numbers are in hexadecimal.

Table 5. Tag Tokens

TAG NAME	TOKEN	TAG NAME	TOKEN
a	1C	postfield	21
anchor	22	pre	1B
access	23	prev	32
b	24	onevent	33
big	25	optgroup	34
br	26	option	35
card	27	refresh	36
do	28	select	37
em	29	setvar	3E
fieldset	2A	small	38
go	2B	strong	39
head	2C	table	1F
i	2D	td	1D
img	2E	template	3B
input	2F	timer	3C
meta	30	tr	1E
noop	31	u	3D
p	20	wml	3F

14.3.3 Attribute Start Tokens

The following token codes represent the start of an attribute in code page zero (0). All numbers are in hexadecimal.

Table 6. Attribute Start Tokens

ATTRIBUTE NAME	ATTRIBUTE VALUE PREFIX	TOKEN	ATTRIBUTE NAME	ATTRIBUTE VALUE PREFIX	TOKEN
accept-charset		5	iname		16
accesskey		5E	label		18
align		52	localsrc		19
align	bottom	6	maxlength		1A
align	center	7	method	get	1B
align	left	8	method	post	1C
align	middle	9	mode	nowrap	1D
align	right	A	mode	wrap	1E
align	top	B	multiple	false	1F
alt		C	multiple	true	20
class		54	name		21
columns		53	newcontext	false	22
content		D	newcontext	true	23
content	application/vnd.wap.wmlc;charset=	5C	onenterbackward		25
			onenterforward		26
domain		F	onpick		24
emptyok	false	10	ontimer		27
emptyok	true	11	optional	false	28
enctype		5F	optional	true	29
enctype	application/x-www-form-urlencoded	60	path		2A
			scheme		2E
enctype	multipart/form-data	61	sendreferer	false	2F
			sendreferer	true	30
format		12	size		31
forua	false	56	src		32
forua	true	57	src	http://	58
height		13	src	https://	59
href		4A	ordered	true	33
href	http://	4B	ordered	false	34
href	https://	4C	tabindex		35
hspace		14	title		36
http-equiv		5A	type		37
http-equiv	Content-Type	5B	type	accept	38
http-equiv	Expires	5D	type	delete	39
id		55	type	help	3A
ivalue		15			

ATTRIBUTE NAME	ATTRIBUTE VALUE PREFIX	TOKEN	ATTRIBUTE NAME	ATTRIBUTE VALUE PREFIX	TOKEN
type	password	3B	type	reset	47
type	onpick	3C	type	text	48
type	onenterbackward	3D	type	vnd.	49
type	onenterforward	3E	value		4D
type	ontimer	3F	vspace		4E
type	options	45	width		4F
type	prev	46	xml:lang		50

14.3.4 Attribute Value Tokens

The following token codes represent attribute values in code page zero (0). All numbers are in hexadecimal.

Table 7. Attribute Value Tokens

ATTRIBUTE VALUE	TOKEN
.com/	85
.edu/	86
.net/	87
.org/	88
accept	89
bottom	8A
clear	8B
delete	8C
help	8D
http://	8E
http://www.	8F
https://	90
https://www.	91
middle	93
nowrap	94
onenterbackward	96
onenterforward	97
onpick	95
ontimer	98
options	99
password	9A
reset	9B
text	9D
top	9E
unknown	9F
wrap	A0
www.	A1

© Copyright Wireless Application Protocol Forum, Ltd., 1998, 1999
All rights reserved

14.4 WML Encoding Examples

Refer to [WBXML] for additional examples.

The following is another example of a tokenised WML deck. It demonstrates variable encoding, attribute encoding and the use of the string table. Source deck:

```
<wml>
    <card id="abc" ordered="true">
      <p>
        <do type="accept">
<go href="http://xyz.org/s"/>
        </do>
        X: $(X)<br/>
        Y: $(&#x59;)<br/>
        Enter name: <input type="text" name="N"/>
      </p>
    </card>
</wml>
```

Tokenised form (numbers in hexadecimal) follows. This example only uses inline strings and assumes that the character encoding uses a NULL terminated string format. It also assumes that the character encoding is UTF-8:

```
02   08   6A   04   'X'  00   'Y'  00   7F   E7   55   03   'a'  'b'  'c'  00
33   01   60   E8   38   01   AB   4B   03   'x'  'y'  'z'  00   88   03
's'  00   01   01   03   ' '  'X'  ':'  ' '  00   82   00   26   03   ' '  'Y'
':'  ' '  00   82   02   26   03   ' '  'E'  'n'  't'  'e'  'r'  ' '  'n'
'a'  'm'  'e'  ':'  ' '  00   AF   48   21   03   'N'  00   01   01   01   01
```

In an expanded and annotated form:

Table 8. Example Tokenised Deck

TOKEN STREAM	DESCRIPTION
02	WBXML Version number 1.2
08	WML 1.2 Public ID
6A	Charset=UTF-8 (MIBEnum 106)
04	String table length
'X', 00, 'Y', 00	String table
7F	`wml`, with content
E7	`card`, with content and attributes
55	`id=`
03	Inline string follows
'a', 'b', 'c', 00	string
33	`ordered="true"`
01	END (of `card` attribute list)
60	`p`
E8	`do`, with content and attributes
38	`type=accept`
01	END (of do attribute list)

TOKEN STREAM	DESCRIPTION
AB	`go`, with attributes
4B	href="http://"
03	Inline string follows
'x', 'y', 'z', 00	string
88	`".org/"`
03	Inline string follows
's', 00	string
01	END (of `go` element)
01	END (of `do` element)
03	Inline string follows
' ', 'X', ':', ' ', 00	String
82	Direct variable reference (`EXT_T_2`)
00	Variable offset 0
26	`br`
03	Inline string follows
' ', 'Y', ':', ' ', 00	String
82	Direct variable reference (`EXT_T_2`)
02	Variable offset 2
26	`br`
03	Inline string follows
' ', 'E', 'n', 't', 'e', 'r', ' ', 'n', 'a', 'm', 'e', ':', ' ', 00	String
AF	`input`, with attributes
48	`type="text"`
21	`name=`
03	Inline string follows
'N', 00	String
01	END (of `input` attribute list)
01	END (of `p` element)
01	END (of `card` element)
01	END (of `wml` element) 15. Static Conformance Statement

15. Static Conformance Statement

This section defines the static conformance requirements for the WML user agent, documents, and encoder.

15.1 WML User Agent

15.1.1 Character Set and Encoding

ITEM	FUNCTION	REFERENCE	MANDATORY/OPTIONAL
WML-01.	UTF-8 Encoding	0	O
WML-02.	UTF-16 Encoding	0	O
WML-03.	UCS-4 Encoding	0	O
WML-04.	Other character encoding	0	O
WML-05.	Reference processing	0	M
WML-06.	Character entities	0	M

15.1.2 Events and Navigation

ITEM	FUNCTION	REFERENCE	MANDATORY/OPTIONAL
WML-07.	History	0	M
WML-08.	Card/Deck task Shadowing	0	M
WML-09.	Intrinsic Events	0	M

15.1.3 State Model

ITEM	FUNCTION	REFERENCE	MANDATORY/OPTIONAL
WML-10.	Browser context	0	M
WML-11.	Initialisation (newcontext)	0	M
WML-12.	Variables	0	M
WML-13.	Context restrictions	0	M

15.1.4 User Agent Semantics

ITEM	FUNCTION	REFERENCE	MANDATORY/OPTIONAL
WML-14.	Deck access control	0	M
WML-15.	Low-memory behaviour	0	O
WML-16.	Error handling	0	M
WML-17.	Unknown DTD handling	0	M
WML-18.	Inter-card navigation	0	M

15.1.5 Elements

If a user agent does not support an optional element, it should continue to process the children of the element. The children of an element include all elements and character data.

ITEM	FUNCTION	REFERENCE	MANDATORY/OPTIONAL
WML-19.	a	0	M
WML-20.	anchor	0	M
WML-21.	access	0	M
WML-22.	b	0	O
WML-23.	big	0	O
WML-24.	br	0	M

ITEM	FUNCTION	REFERENCE	MANDATORY/OPTIONAL
WML-25.	card	0	M
WML-26.	do	0	M
WML-27.	em	0	O
WML-28.	fieldset	0	O
WML-29.	go	0	M
WML-30.	head	0	M
WML-31.	i	0	O
WML-32.	img	0	M
WML-33.	input	0	M
WML-34.	meta	0	O
WML-35.	noop	0	M
WML-36.	p	0	M
WML-37.	postfield	0	M
WML-38.	pre	0	O
WML-39.	prev	0	M
WML-40.	onevent	0	M
WML-41.	optgroup	0	O
WML-42.	option	0	M
WML-43.	refresh	0	M
WML-44.	select	0	M
WML-45.	small	0	O
WML-46.	strong	0	O
WML-47.	table	0	M
WML-48.	template	0	M
WML-49.	timer	0	M
WML-50.	td	0	M
WML-51.	tr	0	M
WML-52.	u	0	O
WML-53.	setvar	0	M
WML-54.	wml	0	M

15.2 WML Encoder

15.2.1 Token Table

ITEM	FUNCTION	REFERENCE	MANDATORY/OPTIONAL
WML-55.	WML token table	0	M

© Copyright Wireless Application Protocol Forum, Ltd., 1998, 1999
All rights reserved

15.2.2 Validation

ITEM	FUNCTION	REFERENCE	MANDATORY/OPTIONAL
WML-56.	XML Well-formed	0	M
WML-57.	XML Validation	0	O
WML-58.	WML Validation	0	O

15.3 WML Document

ITEM	FUNCTION	REFERENCE	MANDATORY/OPTIONAL
WML-59.	Variable references may only occur in **vdata** attribute values	0	M
WML-60.	Variable references must match the production rule **var**	0	M
WML-61.	Two or more **do** elements with the same **name** must not be present in a single card or in the **template** element. (Note: An unspecified **name** defaults to the value of the **type** attribute.)	0	M
WML-62.	A **meta** element must not contain more than one attribute of **name** and **http-equiv**.	0	M
WML-63.	The number of columns in a table must not be set to zero	0	M
WML-64.	Event bindings must not conflict	0	M

Glossary

3G (third generation) wireless. The next step in the development of wireless communications. The first generation was analog and the second was digital (CDMA, TDMA and GSM). Third generation systems are expected to provide broadband, high-speed data applications—both fixed and mobile.

access fee. A special fee that local telephone companies are allowed to charge all telephone customers for the right to connect with the local telephone network.

acknowledgment. The transmission of a short packet from the receiving device to the sending device to indicate that the data sent has been received error-free.

air time. The actual time spent talking on the wireless telephone. Most carriers bill customers based on how many minutes of air time they use each month.

alphanumeric. A message or other type of readout containing both letters ("alphas") and numbers ("numerics").

AMPS (Advanced Mobile Phone System). An analog mobile telephone network that is used mainly in the United States but also Latin America, Australia, New Zealand, parts of Russia and Asia-Pacific.

analog. The original cellular telephone technology still in use today. It allows only one subscriber to use a channel at a time. Also, the traditional method of modulating radio signals so that they can carry information. AM (amplitude modulation) and FM (frequency modulation) are the two most common methods of analog modulation.

antenna. A device for transmitting and receiving signals. The size and shape of antennas are determined, in large part, by the frequency of the signal they are receiving. Antennas are needed on both the wireless handset and the base station.

API (application programming interface). Software used by an application program to request and carry out lower-level services performed by a computer's or telephone system's operating system.

ASCII (American Standard Code for Information Interchange). A standard code used by computer and data communication systems for translating characters, numbers, and punctuation into digital form.

Asynchronous Mode. A standard for data transmission where each data package has a start and stop bit. *See also* synchronous mode.

AT commands. A standardized command set for modems.

authentication. A process used by the wireless carriers to verify the identity of a mobile station. Authentication enables call routing and accurate billing and inhibits unauthorized usage of the system.

bandwidth. A relative range of frequencies that can carry a signal without distortion on a transmission medium.

base station. The central radio transmitter/receiver that maintains communications with a mobile radiotelephone with a given range.

battery. A chargeable device used to provide cellular telephones with power.

Bluetooth. A technology specification for small form factor, low-cost, short range radio links between mobile PCs, mobile phones, and other portable devices.

bps (bits per second). A measure of data transmission speed; the number of pieces of information transmitted per second.

browser. Software that moves documents on the World Wide Web to your computer, PDA, or phone, and displays them. *See also* HDML, HTML, HTTP, WML, and microbrowser.

caller ID. A call-screening feature that lets the user pinpoint the origin of an incoming call prior to answering the phone.

CDMA (Code Division Multiple Access). A spread-spectrum approach to digital transmission pioneered by QUALCOMM. With CDMA, each conversation is digitized and then tagged with a code. The mobile phone is then instructed to decipher only a particular code to pluck the right conversation off the air.

cdmaOne. A family of systems in the wireless communications industry that use QUALCOMM's CDMA digital technology.

CDPD (Cellular Digital Packet Data). Technology for AMPS wireless communications networks that lets data files be broken into a number of 'packets' and sent along idle channels of existing cellular voice networks.

cell. The basic geographic unit of a cellular communications system and the basis for the generic industry term "cellular." A city or county is divided into small cells, each of which is equipped with a low-powered radio transmitter/receiver.

cell site. The location at which communications equipment is located for a cell. A cell site includes antennas, a support structure for those antennas, and communications equipment to connect the site to the rest of the wireless system. cellular systems. Mobile wireless systems that operate at 800 MHz and 1900 MHz frequencies.

channel. A path along which a communications signal is transmitted.

circuit switched. A switching technique that establishes a dedicated and uninterrupted connection between the sender and the receiver.

client/server. A computer network system in which programs and information reside on the server and clients connect to the server for network access.

coverage. The geographical reach of a mobile phone network or system.

D-AMPS (Digital Advanced Mobile Phone Service). A wireless communications standard, also known as IS-136 TDMA. D-AMPS is a digital mobile telephone network that operates in the United States, Latin America, New Zealand, parts of Russia, and Asia Pacific.

DCS 1800. Digital Communications System based on GSM, working on a radio frequency of 1800 MHz. Also known as GSM 1800 and PCN, this digital network operates in Europe and Asia Pacific.

digital. A process whereby information—your speech, for example—is encoded into a stream of zeros and ones before transmission. Digital technologies offer multiple access to radio spectrum—several subscribers can access the same channel at a time. Voice quality and capacity differs markedly among the various digital technologies.

display size. The number of lines and characters, or the number of individual dots or pixels, that can appear on the display of a mobile phone.

DTMF (dual tone multi frequency signal). The audible signals generated by a telephone when you press a key on its keypad.

dual band. Dual band mobile phones can work on networks that operate on different frequency bands.

dual mode. Dual mode mobile phones work on both digital and analog wireless networks.

electromagnetic spectrum. The full range of frequencies, from the very lowest (zero cycles per second) to the very highest (just less than an infinite number of cycles per second). *See also* frequency and hertz.

e-mail (electronic mail). Messages sent across communications networks—both wireless and landline.

encryption. The transformation of data, for the purpose of privacy, into an unreadable format until reformatted with a decryption key.

EPOC. An operating system, designed especially for mobile use, being put forward by Ericsson, Nokia, Psion, Motorola, and Matsushita.

ergonomics. The study of equipment design in order to reduce user fatigue and discomfort.

ESMR (Enhanced Specialized Mobile Radio). Digital mobile telephone services offered to the public over channels previously used for two-way analog dispatch services. *See also* SMR.

ESN (electronic serial number). The unique number assigned to a wireless phone by the manufacturer. According to the Federal Communications Commission, the ESN is to be fixed and unchangeable—a unique fingerprint for each phone. *See also* MIN.

ETACS (Extended Total Access Communications System). The analog mobile phone network developed in the UK and available in Europe and Asia.

ETSI. The European Telecommunications Standards Institute.

extranet. An Intranet-like network which a company extends to conduct business with its customers and/or its suppliers. Extranets generally have secure areas to provide information to customers and external partners.

FCC (Federal Communications Commission). The government agency responsible for regulating telecommunications in the United States, located in Washington, D.C.

FDMA (frequency division multiple access). A method of radio transmission that allows multiple users to access a group of radio frequency bands without interference.

frequency. A measure of the energy, as one or more waves per second, in an electrical or light-wave information signal. A signal's frequency is stated in either cycles-per-second or hertz (Hz). *See also* hertz.

frequency reuse. The ability to use the same frequencies repeatedly across a cellular system, made possible by the basic design approach for cellular. Since each cell is designed to use radio frequencies only within its boundaries, the same frequencies can be reused in other cells.

full duplex. A communications connection in which both parties can talk at the same time. The opposite is simplex functionality, when one party talks and the other must wait for his/her turn.

gateway. An Internet-based server that acts as an intermediary for another server.

GHz (gigahertz, billions of hertz). A measure of spectrum corresponding to a billion cycles per second. Personal Communications Services operate in the 1.9 GHz band of the electromagnetic spectrum. *See also* hertz, kHz, and MHz.

GPRS (General Packet Radio Service). An extension to the GSM standard to include packet data services.

GPS (Global Positioning System). A satellite system using 24 satellites orbiting the Earth from 10,900 miles away that lets users pinpoint precise locations using the satellites as reference points.

GSM (Global System for Mobile Communications). GSM usually refers to the European standard GSM operating on 900 MHz and 1800 MHz bands, but in North America it refers to the 1900 MHz band.

GSM 1800. Also known as DCS 1800 or PCN, GSM 1800 is a digital network working on a frequency of 1800 MHz. It is used in Europe, Asia Pacific, and Australia.

GSM 1900. Also known as PCS 1900, GSM 1900 is a digital network working on a frequency of 1900 MHz. It is used in the US and Canada and is scheduled for parts of Latin America and Africa.

GSM 900. GSM networks operating in the 900 MHz band, predominantly in Europe.

GUI (graphical user interface). Any computer interface that substitutes graphics for characters.

handoff. The process by which a mobile telephone switching office passes a cellular phone conversation from one radio frequency in one cell to another radio frequency in another call.

HDML (Handheld Device Markup Language). A modification of standard HTML, developed by Phone.com, for use on the small screens of mobile phones, PDAs, and pagers.

HDML is a text-based markup language that uses Hypertext Transfer Protocol (HTTP) and is compatible with web servers.

HDR (high data rate). High-speed, high-capacity wireless technology that provides up to 2.4 Mbps in a standard bandwidth 1.25 MHz channel.

hertz. A measurement of electromagnetic energy, equivalent to one wave or cycle per second. *See also* kHz, MHz, and GHz.

HSCSD (high speed circuit switched data). A feature of certain GSM networks that lets a digital telephone communicate at up to 40 kHz for data communications.

HTML (Hypertext Markup Language). An authoring software language used on the web. HTML is used to create web pages and hyperlinks.

HTTP (Hypertext Transfer Protocol). The protocol used by the web server and the client browser to communicate and move documents around the Internet.

iDEN (Integrated Dispatch Enhanced Network). A wireless technology developed by Motorola that works in the 800 MHz, 900 MHz, and 1.5 GHz radio bands. The technology supports, on one handset, voice-both dispatch radio, numeric paging, Short Message Service (SMS), data and fax transmission.

IMT-2000 (International Mobile Telecommunications-2000). The standard for third generation mobile communications systems. In Europe, it is called UMTS and in Japan it is called J-FPLMTS.

infrared. A band of the electromagnetic spectrum used for airwave communications and some fiber-optic transmission systems. Infrared is commonly used for short-range (up to 20 feet) through-the-air data transmission. Many PC devices have infrared ports, called IrDA ports.

infrastructure. All parts of the wireless network, excluding the subscriber handset. It includes the switching hardware, base stations, cell sites, and all the links between them.

intelligent roaming. Intelligent roaming registers PCS mobile telephones with acceptable service providers, ensuring customers the best possible service at any location.

international roaming. A mobile telephone feature that lets you switch between networks, offering coverage abroad.

Internet. A global network of linked computer networks, popularized by a graphical interface called the World Wide Web.

intranet. An internal network that is private or employs a firewall to secure it from outside access, and supports Internet technology. An intranet is used for inter-company communications and can be accessed only by authorized users.

IrDA (Infrared Data Association). A standard for wireless connectivity that uses infrared light to communicate between computing devices.

IS-136 TDMA. A United States cellular standard also known as D-AMPS.

Java. A programming language from Sun Microsystems which abstracts programs into bytecodes so that the same software runs on any operating system.

J-FPLMTS (Japanese Future Public Land Mobile Telecommunications Services). The Japanese equivalent of the IMT-2000 third generation technology standard.

kHz (kilohertz, thousands of hertz). A measure of frequency corresponding to one thousand hertz or cycles per second. *See also* hertz, MHz, and GHz.

menu. User interface for changing the settings on computer and mobile telephones.

MHz (megahertz). A measure of frequency corresponding to one million hertz or cycles per second. Most wireless communications systems operate in the 800 and 900 MHz bands of the electromagnetic spectrum. *See also* hertz, kHz, and GHz.

microbrowser. Software that moves documents on the World Wide Web to your computer, PDA, or telephone and displays them. Microbrowsers are designed for devices with minimal memory, processing power, and small display screens.

MIN (mobile identification number). A number assigned by the wireless carrier to a customer telephone. The MIN is meant to be changeable, since the phone can change hands or a customer can move to another city.

modem (modulator/demodulator). A hardware device which converts digital data into analog and vice versa to enable digital signals from computers to be transmitted over analog telephone lines.

MTSO (mobile telephone switching office). The central switch that controls the entire operation of a cellular system. It is a sophisticated computer that monitors all cellular calls and tracks the location of all cellular-equipped vehicles traveling in the system.

NAM (Number Assignment Module). The electronic memory in the cellular phone that stores the telephone number. Phones with dual- or multi-NAM features offer users the option of registering the phone with a local number in more than one market.

network. A cellular telephone system consisting of a network of cells. Each cell is served by a radio base station from which calls are forwarded to and received from your mobile phone by wireless radio signals.

NiCd (nickel cadmium). A type of battery suitable for general use. NiCd batteries are robust and long-lasting.

NiMH (nickel metal hydride). A type of battery especially good for heavy use. NiMH batteries hold more power for their size than NiCd batteries as they always accept a full charge. MiMH batteries are also more environmentally friendly.

NMTN (Nordic Mobile Telephone Network). The first cellular system in the world to come into service. NMTN is an analog network used in Scandinavia, some European countries, and small parts of Russia, the Middle East, and Asia.

numeric. Numeric means numbers. A numeric pager, for example, can send and receive messages containing only numbers. The alternative is alphanumeric, meaning both letters and numbers.

off-peak. The period of time after the business day ends during which wireless carriers may offer reduced airtime charges.

operating system. A software program which manages the basic operations of a computer system. These operations include memory apportionment, the order and method of handling tasks, flow of information into and out of the main processor and to peripherals, and other maintenance-level tasks.

packet. *See* packet data.

packet data. A bundle of data organized in a specific way for transmission. The three principal elements of a packet include the header, the text, and the trailer (error detection and correction bits).

packet radio. The transmission of data over radio using a version of the X.25 data communications protocol. The data is broken into packets and transmitted wirelessly.

packet switching. Sending data in packets through a network to a remote location. The data is assembled by a modem into individual packets of data.

pager. Small portable message receivers that are generally inexpensive, reliable, and have nationwide coverage. Pagers began as one-way devices, but two-way paging capabilities are available over some networks, notably packet data and PCS networks.

PCN (Personal Communications Network). Also known as the DCS 1800 standard or GSM 1800. A wireless communications network used in Europe and Asia Pacific.

PCS (Personal Communications Service). FCC terminology describing two-way, personal, digital wireless communications systems. Wireless communications networks operating in the 1.9 GHz band in North America. It includes three technologies: GSM 1900, CDMA, and TDMA (IS-136, or D-AMPS).

PCS 1900. *See* PCS.

PDA (personal digital assistant). Portable computing devices capable of transmitting data. These devices make possible services such as paging, data messaging, electronic mail, stock quotations, handwriting recognition, personal computing, facsimile, date book, and other information handling.

PDC (Personal Digital Cellular). The digital network used primarily in Japan.

peak. Highest-usage period of the business day when a cellular system carries the most calling traffic.

PIM (personal information manager). One or more programs that log personal and business information such as contacts, appointments, lists, notes, events, and memos.

PIN (personal identification number). A code used for all GSM-based phones to establish authorization for access to certain functions or information.

predictive text input. A data entry method in which a computer program tries to predict the word you are entering based on just a few characters.

protocol. A specific set of rules for organizing the transmission of data in a network.

proxy server. An Internet-based server that intermediates between a browser (or microbrowser) and a server.

quick charge. A charger feature that lets your cellular telephone battery charge in less than one hour.

retractable antenna. A telescopic antenna which extends and retracts into the telephone housing.

RF (radio frequency). A frequency well above the range of human hearing.

roaming. The ability to use your cellular phone outside your usual service area, when traveling, for example.

server. Any Internet-connected computer that is capable of delivering documents to another Internet-connected computer for viewing or other operation.

service plan. A rate plan selected by subscribers when they start up cellular service, usually consisting of a base rate for system access and a per-minute rate for usage.

service provider. A company that provides services and subscriptions to mobile phone users.

Short Message Service (SMS). A feature available on digital networks allowing messages of up to 160 characters to be sent and received, via the network operator's message center, by your mobile phone.

SIM (Subscriber Identity Module). A computer chip set in a GSM handset that contains information needed to identify the subscriber when connecting to the network, especially for billing purposes.

smart charge sensor. A battery charger feature that prevents your cellular telephone's battery from being overcharged.

smart phone. A phone with a microprocessor, memory, screen, a built-in modem, and possibly a keyboard. A smart phone combines some of the capabilities of a PC with a cellular telephone handset.

SMR (Specialized Mobile Radio). A U.S.-only private business service that uses mobile radio telephones and base stations similar to other wireless services. It is usually used in dispatch applications, such as delivery companies or taxicab organizations.

spread spectrum. A modulation technique, also known as frequency hopping, used in wireless systems. The data is packetized and spread over a range of bandwidth.

standby time. The amount of time you can leave your fully charged cellular telephone turned on before it completely discharges the batteries. *See also* talk time.

synchronization. The process of uploading and downloading information from two or more databases, so that each is identical.

TACS (Total Access Communication System). The 900 MHz British analog mobile telephone standard based on the U.S. AMPS system.

talk time. The length of time you can talk on your portable or transportable cellular telephone without recharging the battery. *See also* standby time.

TCP/IP (Transmission Control Protocol/Internet Protocol). The standard set of protocols used by the Internet for transferring information between computers, handsets, and other devices.

TDMA (time division multiple access). A method of digital wireless communications transmission allowing a large number of users to access (in sequence) a single radio frequency channel without interference by allocating unique time slots to each user within each channel. A digital mobile phone network that operates in the United States, Latin America, New Zealand, parts of Russia and Asia Pacific. Also known as IS-136 and D-AMPS.

third generation. *See* 3G.

transport layer. The lower levels of a wireless communications system that are responsible for the basic movement of signals from place to place.

trickle charge. A battery charger feature that prevents loss of battery capacity. The charger senses when the battery is fully charged and switches to trickle mode, ensuring that the battery has full capacity when removed.

triple mode. A combined analog and digital mobile phone that operates in the existing analog system frequency (800 MHz) and in both digital frequencies (800 MHz and 1900 MHz).

UMTS (Universal Mobile Telecommunications Services). The European term for wireless systems based on the IMT-2000 standard.

vocoder. A device used to convert speech into digital signals.

voice mail. A computerized answering service that automatically answers your call, plays a greeting in your own voice, and records a message. After you retrieve your messages, you can delete, save, reply to, or forward the messages to someone else. Also called voice messaging.

voice messaging. *See* voice mail.

voice-activated dialing. A feature that lets you dial a telephone number by speaking to a wireless phone instead of using a keypad.

voice-operated transmitter (VOX). A cellular telephone feature that conserves your battery's power by transmitting only when you are speaking.

WAP (Wireless Application Protocol). A protocol for creating wireless applications that let cellular telephone and mobile device users access Internet-based content and services.

WCDMA (wideband CDMA). A "third generation" (or 3G) wireless service that will expand the range of options available to users and allow communication, information and entertainment services to be delivered at much higher speeds than today's wireless networks.

web. *See* World Wide Web.

wireless. Radio-based systems that allow transmission of telephone and/or data signals through the air without a physical connection such as a metal wire or fiber optic cable.

Wireless Application Protocol (WAP). *See also* WAP.

WML (Wireless Markup Language). A programming language for creating WAP applications that run on wireless telephones and other mobile devices. *See also* HDML.

WMLScript. A lightweight scripting language used to add sophisticated computational capabilities to WAP applications programs.

World Wide Web. A collection of computers throughout the world that are capable of delivering content and services to browser-equipped computers and devices.

X.25. An international protocol standard for packet switched networks.

XML. An international standard for defining programming and content-delivery languages for use on the Internet.

Index

3G. *See* Third generation.
4thpass KBrowser, 64
123Jump, 73

A

A (element), 136–137
accept (event), 54
Access (element), 145–146
Access fee, 191
Acknowledgment, 191
Acrobat Reader (Adobe), 34
Advanced Mobile Phone
    Service (AMPS), 2, 8–9, 191
  networks, 11
African Cellular, 105
Air time, 191
Allaire, 19
AllNetDevices, 105
Alphanumeric, 87, 89, 98, 191
American Standard Code for
    Information Interchange
    (ASCII), 140, 192
AMPS. *See* Advanced Mobile
    Phone Service.
Analog, 191
  communication, digital
      communication
      contrast, 4–5

Anchor. *See* Fragment anchors.
    (element), 135–136
Anchors. *See* HyperText
    Markup Language;
    Wireless Markup Language.
Antenna, 191. *See also*
    Retractable antenna.
Anywhereyougo.com, 105
API. *See* Application
    Programming Interface.
Apple Newton PDA, 57
Application Designer. *See*
    Ericsson.
Application layer, 7
Application Programming
    Interface (API), 38, 78, 192
ASCII. *See* American Standard
    Code for Information
    Interchange.
Asynchronous device, 161
Asynchronous mode, 192
AT commands, 192
AT&T, 15, 23
  Wireless, 14, 18
Attribute
  start tokens, 183–184
  tokens, 180
  value tokens, 184

Attributes, 46–47, 122–123,
    128–132, 134–139
  names, encoding, 181
AU-System, 59–61
Authentication, 39, 79, 192. *See
    also* Client authentication.
Author-specified URL, 126

B

Backus-Naur Form (BNF), 118.
    See also Extended Backus-
    Naur Form.
Bandwidth, 27, 192
Base station, 192
Battery, 10, 26, 66, 99, 100,
    101, 192
BBC News, 73
Bearers, 39
Beginning tag, 46
Bell Atlantic, 14, 20
Bell South Wireless, 18, 22
Bertelsmann AG/America
    Online, 20
Binary data encoding, 76–77
Binary encoding. *See* Wireless
    Application Protocol;
    Wireless Markup
    Language.

B

Binary representation. *See* Wireless Markup Language.
Bits per second (bps), 10, 11, 12, 95, 192
Bluetooth, 192
BNF. *See* Backus-Naur Form.
Boolean type, 125
Bps. *See* Bits per second.
Br (element), 165
Br (tag), 50
Browser, 192. *See also* Desktop browser; Ericsson; Microbrowser; Original Equipment Manufacturer browser; Phone.com.
  context, 139, 170, 172
  size. *See* Limited browser context size.

C

Calendaring, 87
Caller ID, 3, 192
Card (element), 148–150
Card/deck task shadowing, 132–133
Card intrinsic events, 148
Cards, 47–48, 115, 149. *See also* Destination card.
Case sensitivity, 123
CDATA section, 123–124
CDMA. *See* Code Division Multiple Access.
cdmaOne, 192
CDPD. *See* Cellular Digital Packet Data.
Cell, 192
  site, 192
Cellular Digital Packet Data (CDPD), 6, 11, 70, 85, 192
Cellular service providers, 18
Cellular Telecommunications Industry Association (CTIA), 107
Cellular telephone manufacturers, 26
CGI. *See* Common Gateway Interface.
Channel, 4, 9, 10, 192
Character
  data, 124
  encoding, 117
  entities, 121–122
  set. *See* Wireless Markup Language.
Circuit switched, 193
  connections, packet data connections contrast, 3–4
  data. *See* High speed circuit switched data.
  networks, 95
class (attribute), 125–126
Client authentication, 79
Client/server, 193
Code Division Multiple Access (CDMA), 9–10, 16, 67, 106, 192. *See also* Wideband CDMA.
CDMA2000, 12
CDMA-based services, 87
data modem, 70
Code page, 183, 184
Comments, 123
Common Gateway Interface (CGI), 34
Communications infrastructure, 24
Compact binary representation. *See* Wireless Markup Language.
Computing power, 23–24
Conformance statement. *See* Static conformance statement.
Consumer microbrowsers, 61–64
Consumer profile. *See* Wireless Application Profile.
Consumer WAP sites, 72–74
Content, 49–51
  duplication, 99
  encoding, 117
Content/service providers, 19
ContentType (type), 126
Context. *See* Browser.
  management, features, 43
  restrictions, 143
  size. *See* Limited browser context size.
Control elements, 150–160
Convergence. *See* Wireless Application Protocol.
Core WML data types, 124–126
Coverage, 8, 22, 23, 71, 86, 193
CTIA. *See* Cellular Telecommunications Industry Association.

D

D-AMPS. *See* Digital Advanced Mobile Phone Service.
Data
  connections, 12
  encryption, 79, 80
  entry, 55–56, 98, 100. *See also* Predictive data entry.
  integrity, 79
  link layer, 7
  networks. *See* Voice/data networks.
  types, 58. *See also* Core WML data types.
DCS. *See* Digital Communications System.
DDI, 15
Deck-level element, 132
Decks, 47–48, 115, 132, 139, 145–168
  access control, 170
  structure. *See* Wireless Markup Language.
delete (event), 55
Desktop browser, 98
Destination card, 171, 172
Devices. *See* Wireless Application Protocol.
  types. *See* Wireless Application Protocol Wireless Markup Language (WAP WML).
Dialects, support, 43
Dialogs, 58
Digital, 193
  communication, contrast. *See* Analog.
Digital Advanced Mobile Phone Service (D-AMPS), 9, 70, 85–86, 193
Digital Communications System (DCS) 1800, 193
Directories. *See* Mobile Web.

Display size, 119, 193
dmoz (search engine), 105
Do (element), 133–135, 150
DoCoMo, 15, 18
DOCTYPE, 181
Document Type Definition (DTD), 173–180. *See also* Unknown DTD.
Documents, 103–104. *See also* Wireless Markup Language.
  identifiers, 173
  prologue, 144
  status. *See* Wireless Application Protocol Wireless Markup Language.
  validation, 181
Dollar-sign character, 142
DTD. *See* Document Type Definition; Unknown DTD.
DTMF. *See* Dual Tone Multi Frequency.
Dual band, 193
Dual mode, 193
Dual Tone Multi Frequency (DTMF) signal, 193

E

EBNF. *See* Extended Backus-Naur Form.
ECMAScript, 40, 57
Economic forces, 24–25
EDGE, 12
Electromagnetic spectrum, 2, 7, 193
Electronic mail (E-mail), 74, 87, 93, 107, 193
  exchange, 7
Electronic serial number (ESN), 193
Elements, 46–47, 122, 187–188
E-mail. *See* Electronic mail.
Emphasis, 162–164
Empty-element tag, 122
emptyok (attribute), 55
Encoder. *See* Wireless Markup Language.
Encoding. *See* Attributes; Binary data encoding; Content; Semantics; Tag names; Variables; Wireless Application Protocol.
  examples. *See* Wireless Markup Language.
Encryption, 44, 79, 80, 193
Ending tag, 46
Enhanced Specialized Mobile Radio (ESMR), 193
Enterprise IT managers, 75
Enterprise Resource Planning (ERP), 81, 82
Enterprise WAP strategy
  architecture, 92–93
  design/implementation, 93–96
  implementation, 91
  requirements, 91–92
  testing, 96
Entities, 122
EPOC, 61, 193
  operating system, 68
Ergonomics, 193
Ericsson, 14, 16, 18, 20, 23, 78, 105
  Application Designer, 29
  MC218, 45, 60, 68–69
  phone, 32
  R280, 85–86
  R320, 68–69, 83, 84
  R380, 68–69
  RS320/RS380, 60
  WAP Browser, 60
ERP. *See* Enterprise Resource Planning.
Errors, 124
  handling, 171
ESMR. *See* Enhanced Specialized Mobile Radio.
ESN. *See* Electronic serial number.
ETACS. *See* Extended Total Access Communications System.
ETSI. *See* European Telecommunications Standards Institute.
European Analog Cellular, 8–9
European Telecommunications Standards Institute (ETSI), 193
Event-handling elements, 132
Events, 51–55, 126–139, 187. *See also* Intrinsic events.
  handling, 126
Excite, 73
Execution failure. *See* Tasks.
exo-net, 81–83
  business background, 81–82
  features, 82
  project status, 83
  technology/development, 83
  WAP background, 82–83
Extended Backus-Naur Form (EBNF), 140
Extended Total Access Communications System (ETACS), 193
Extensible Markup Language (XML), 41, 46, 93, 119–125, 140, 162, 199. *See also* Wireless Binary XML.
  application/processor, 142
  Content Format. *See* Wireless Binary XML.
  data types, 124
  declaration, 144
Extension token assignment. *See* Wireless Markup Language.
Extension tokens, 180. *See also* Global extension tokens.
Extranet, 194

F

FCC. *See* Federal Communication Commission.
FDMA. *See* Frequency Division Multiple Access.
Federal Communication Commission (FCC), 2, 3, 194
Fieldset (element), 159–160
  examples, 160
File transfer, 7
File Transfer Protocol (FTP), 33
Firewall, 76
Flow type, 125
format (attribute), 55, 158
Fragment anchors, 120

Frequencies, 194
  explanation, 2–3
  reuse, 194
Frequency Division Multiple Access (FDMA), 194
FTP. *See* File Transfer Protocol.
Full duplex, 194

G
Gaddo.net, 105
Gateway, 35, 75–80, 194. *See also* Wireless Application Protocol.
  finding, 78
  interface. *See* Common Gateway Interface.
  security, 79
  software extensions, 78
Gelon.net, 105
General Packet Radio Service (GPRS), 12, 194
GHz. *See* Gigahertz.
GIF, 51
Gigahertz (GHz), 2, 194
Global extension tokens, 180
Global Positioning System (GPS), 88, 194
Global System for Mobile Communications (GSM), 9, 194
  900, 65, 66, 68, 69, 194
  1800, 65, 66, 68, 69, 194
  1900, 12, 194
Global System for Mobile Telecommunications (GSM), 10
  GSMData, 105–106
Glossary, 191–199
Go (element), 128–131
Go (task), 171–172
GPRS. *See* General Packet Radio Service.
GPS. *See* Global Positioning System.
Graphical User Interface (GUI), 194
GSM. *See* Global System for Mobile Communications; Global System for Mobile Telecommunications.

GSMData. *See* Global System for Mobile Telecommunications.
GTE, 14, 20
GUI. *See* Graphical User Interface.

H
Handheld Device Markup Language (HDML), 14, 60, 77, 86, 194–195
  services, 19, 88
  technology, 19
Handheld Device Transport Protocol (HDTP), 14
Handheld Markup Language (HDML), 40, 118
Handoff, 194
Handwriting recognition, 57
Hardware. *See* Wireless Application Protocol.
  providers, 18
HDML. *See* Handheld Device Markup Language; Handheld Markup Language.
HDR. *See* High data rate.
HDTP. *See* Handheld Device Transport Protocol.
Head (element), 145–147
help (event), 55
Hertz, 2, 195
High data rate (HDR), 195
High speed circuit switched data (HSCSD), 65, 195
Hitachi, 60
Hole Sun. *See* Wireless Application Protocol.
Holliday Group, 81, 83, 95
Hotlist. *See* Mobile Web.
HREF, 141, 181
  type, 125
HSCSD. *See* High speed circuit switched data.
HTML. *See* HyperText Markup Language.
HTTP. *See* HyperText Transfer Protocol.
HTTPS. *See* HyperText Transfer Protocol Secure.

HyperText Markup Language (HTML), 14, 17, 27, 36, 40–42, 46, 57, 118, 170, 195
  anchors, 54
  content, 92
  data, 76
  files, 34, 92
  HTML-related languages, 99
  pages, 18, 34, 47, 115
  standard, 120
  text, 38
  translation. *See* Wireless Markup Language.
HyperText Transfer Protocol (HTTP), 17, 33, 40, 118, 195
  GET request, 130
  header, 146, 147
  messages, 35
  post method, 127
  POST request, 130
  protocols, 41, 76
  response, 36
  server, 118
  submission method, 128
HyperText Transfer Protocol Secure (HTTPS), 33

I
IANA. *See* Internet Assigned Number Authority.
ICQ standard, 74
id (attribute), 125–126
iDEN. *See* Integrated Dispatch Enhanced Network.
IDO, 15, 18
IIS, 83
Images, 51, 168–170
  support, 42
IMC, 40
i-Mode (Japan), 15–16
Implementation-dependent state, 139
IMT. *See* International Mobile Telecommunications.
Information Technology (IT) managers, 91. *See also* Enterprise IT managers.
Informative references. *See* Wireless Application

Protocol Wireless Markup Language.
Infrared, 68, 69, 195
Infrared Data Association (IrDA), 195
Infrastructure, 14, 16, 17, 24, 81, 86, 106, 195
Input. *See* Predictive input.
  alternatives, 56–57
  devices, 119
  (element), 156–159
    examples, 159
  validation, 44
Integrated Dispatch Enhanced Network (iDEN), 9, 67, 195
Integrity. *See* Data.
Intelligent roaming, 195
Intelligent Terminal Transport Protocol (ITTP), 14
Interactive television, 93
Inter-card linking, 115
Inter-card navigation, 115, 171–173
International Mobile Telecommunications (IMT) 2000, 195
International roaming, 195
International Standards Organization (ISO)
  ISO10646, 116, 120
  network model, 6–8, 37
Internet, 20–22, 24, 61, 68, 74, 195
  access, 101
  connection, 98
  Internet-hosted content, 101
  server, 76
  technologies, 17, 26
Internet Assigned Number Authority (IANA), 118
  Character Set registry, 129
Internet Explorer, 18, 63
Internet Service Provider (ISP), 4
Inter-word space, 164
Intranet, 195
  web server, 78
Intrinsic events, 137–139
IrDA. *See* Infrared Data Association.
IS-136 TDMA, 195

ISBN number, 99
ISO. *See* International Standards Organization.
ISP. *See* Internet Service Provider.
IT. *See* Information Technology.
ITTP. *See* Intelligent Terminal Transport Protocol.

J

Japanese Future Public Land Mobile Telecommunications Services (J-FPLMTS), 195
Java, 14, 17, 64, 77, 195
Java API, 78
Java KVM, 64
JavaScript, 14, 17, 36, 40, 57, 118
J-FPLMTS. *See* Japanese Future Public Land Mobile Telecommunications Services.
JPEG, 51

K

KBrowser. *See* 4thpass KBrowser.
Key Performance Indicators, 82
kHz. *See* Kilohertz.
Kilohertz (kHz), 2, 195
Kyocera, 64

L

Languages, support, 43
Latency, 98–99, 101
Le Hors, Arnaud, 104
%length, validation, 181
Length type, 124
Library functions, 58
Limited browser context size, 170
Limited history. *See* User agent.
Line-wrap mode, 165
Loops, 58
Low-memory behavior, 170

M

MainFreight, 84–85
  business background, 84

project status, 85
  technology/development, 84–85
  WAP background, 84
Man-machine interface (MMI), 118
maxlength (attribute), 55
MCI WorldCom, 23
Media Type. *See* Wireless Markup Language.
Megahertz (MHz), 2, 196
Menu, 30, 53, 87, 115, 196
Messaging. *See* Mobile Web; Voice.
Meta (element), 146–147
Meta-data, 146
Meta-information, 146
MHz. *See* Megahertz.
Microbrowser, 40–41, 45–46, 94, 98, 196. *See also* Consumer microbrowsers; Original Equipment Manufacturer.
  implementation, demands, 96
MIME. *See* Multipurpose Internet Mail Extension.
MIN. *See* Mobile Identification Number.
Mitsubishi T250, 70
MMI. *See* Man-machine interface.
Mobile Access Phone, 14
Mobile Explorer, 60–61
Mobile Identification Number (MIN), 196
Mobile Telephone Switching Office (MTSO), 196
Mobile Web (Verizon)
  bookmarks, 88
  customer care, 88
  directories, 87–88
  hotlist, 87
  messaging, 87
  My.Airtouch, 87
  services, 87–88
  shopping, 88
MobileAccess T250, 70
mobileID, 106
Modem. *See* Modulator/demodulator.

Index

Modulator/demodulator (modem), 4, 5, 8, 10, 11, 15, 22, 26, 61, 65, 67, 70, 196
Moore, Gordon, 23
Motorola, 13, 14, 16, 18, 60, 67
   i500 Plus/i700 Plus, 67
   i1000 Plus, 32, 67
MSN Mobile Services, 87
MTSO. *See* Mobile Telephone Switching Office.
Multipurpose Internet Mail Extension (MIME), 116, 121

N

NAM. *See* Number Assignment Module.
name (attribute), 161
Narrowband network connection, 115, 119
NA-TDMA. *See* North American TDMA.
Navigation, 98, 100, 126–139, 187. *See also* Inter-card navigation.
   history, 126–127
   mechanisms, 43
Navigational accelerator, 151
Navigational user interface, 136
NeoPoint, 60
NeoPoint 1000/1600, 71
Nested elements, 125
Netscape Communicator, 18, 63
Network, 196
   layer, 7
   model. *See* International Standards Organization network model.
Newcontext (attribute), 140
NiCD. *See* Nickel cadmium.
Nickel cadmium (NiCD), 196
Nickel metal hydride (NiMH), 196
NiMH. *See* Nickel metal hydride.
Nippon Telephone and Telegraph (NTT), 15, 18, 23
NMTN. *See* Nordic Mobile Telephone Network.
NMTOKEN, 124, 181
Nokia, 13, 14, 16, 18, 20, 29, 78, 106
6210/6250, 65
7110, 66, 83, 84
Non-formatting characters, 158
Non-repudiation, 80
Non-zero width gutter, 166
Noop
   (element), 131–132
   (task), 132, 172
Nordic Mobile Telephone Network (NMTN), 196
Normative references. *See* Wireless Application Protocol Wireless Markup Language.
North American TDMA (NA-TDMA), 9
NTT. *See* Nippon Telephone and Telegraph.
NULL terminated string format, 185
Number Assignment Module (NAM), 196
Number type, 125
Numeric, 55, 57, 113, 119, 181, 196
Numeric constants, 181–184

O

OEM. *See* Original Equipment Manufacturer.
Off-peak, 196
Onenterbackward (event), 138
Onenterforward (event), 137, 148
On-screen button, 43
Ontimer (event), 137
Open Directory Project, 105
Open Systems Interconnection (OSI) network model, 37
Operating system, 22, 24, 68, 69, 196
Optgroup (element), 154–155
Option (element), 153–154
options (event), 55
Oracle, 19
Original Equipment Manufacturer (OEM) microbrowsers, 59–61
Orktopas, 74
OSI. *See* Open Systems Interconnection.

P

Packet, 196
   data, 197
      connections, contrast. *See* Circuit switched.
   radio, 197
   switching, 197
PageNet, 18
Pager, 39, 197. *See also* Two-way pagers.
Paging. *See* Two-way paging.
Palm Computing, 18, 62
Palm OS, 61
Palm.net, 22
Paoli, Jean, 103
Paragraphs, 164–165
Parsing. *See* Variable substitution syntax.
Password (text entry control), 157
PCDATA, 123, 124
PCN. *See* Personal Communications Network.
PCS. *See* Personal Communications Service; Personal Communications Systems.
PCS Data, 106
PDA. *See* Personal Digital Assistant.
PDC. *See* Personal Digital Cellular.
Peak, 197
Personal Communications Network (PCN), 10, 197
Personal Communications Service (PCS), 197
Personal Communications Systems (PCS), 3, 9
Personal Digital Assistant (PDA), 118, 119
   PDA-class device, 149
Personal Digital Cellular (PDC), 44, 197
Personal identification number (PIN), 197
Personal information manager (PIM), 197
PhoneAFact.com, 73

Index

Phone.com, 18, 22, 29, 77, 78, 106
   UP.Browser, 14, 60, 71, 87
   UP.Simulator, 32
Physical layer, 7
PIM. *See* Personal information manager.
PIN. *See* Personal identification number.
Pop operation, 131
Postfield (element), 127
Pre (element), 168
Predictive data entry, 100
Predictive input, 57
Predictive text input, 197
Presentation layer, 7
Prev
   (element), 131
   (event), 54
   (task), 172
Privacy, maintenance, 79
Processing instructions, 124
Protocol, 5–6, 197
   explanation, 34
   stack, 8
   translation, 76
Proxy server, 76, 197

Q

QUALCOMM, 10, 14, 18, 60, 106
QCP-860, 87
Quick charge, 197

R

Radio frequency (RF), 2, 197
Reference. *See* Wireless Application Protocol Wireless Markup Language.
   information. *See* Wireless Markup Language.
   processing
     behavior, 171–172
     model, 121
Refresh
   (element), 131
   (task), 172
Relative URLs, 120
Request For Comment (RFC), 118
1766, 116, 125
2045, 116, 126, 134
2047, 116
2048, 116
2068, 116, 118, 128, 146
2119, 116
2388, 116, 129
2396, 115, 116, 119, 120, 130, 141
Request-response process, 36
Reserved words, 180
reset (event), 55
Resource-limited devices, 93
Resources, 103
   URLs, 103–104
Retractable antenna, 197
Reverse Lookup, 87
REX platform, 61
RF. *See* Radio frequency.
RFC. *See* Request For Comment.
Roaming, 2, 197. *See also* Intelligent roaming; International roaming.
RSA security, 87

S

Screen size, 97, 99–100
SDK. *See* Software Development Kit.
Search engines. *See* dmoz; Wireless Application Protocol.
Secure Sockets Layer (SSL), 80
Security, 79–80
   contribution. *See* Wireless Application Protocol.
Select (element), 151–153
Select lists, 151–156
   examples, 155–156
Semantics. *See* User agent.
   encoding, 181
Server, 197. *See also* Client/server.
   defined, 75
Session layer, 7
Setup, 4
Setvar (element), 127–128
SGML. *See* Standardized Generalized Markup Language.
SGML Public Identifier, 173
Shadowing. *See* Card/deck task shadowing.
Short Message Service (SMS), 9, 11, 39, 70, 198
Shovelware, 88
SIM. *See* Subscriber Identity Module.
Sky City Hotels, 85–86
   business background, 85
   retrospect, 86
   technology/development, 86
   WAP background, 85–86
Slob-Trot software, 63
Smart charge sensor, 198
Smart Messaging protocol, 14
Smart phone, 15, 18, 19, 21, 23, 24, 27, 35, 44, 45, 54, 57, 62, 72, 82, 83, 87, 198
Smart phone-based services, 15
Smith, Randall B., 104
SMR. *See* Specialized Mobile Radio.
SMS. *See* Short Message Service.
Social forces, 24–25
Software developers, 18–19
Software Development Kit (SDK), 29, 32, 58, 61
   downloading, 62
Sony, 18, 60, 64
Specialized Mobile Radio (SMR), 9, 198
Spectrum. *See* Electromagnetic spectrum; Spread spectrum.
   description, 2–3
Sperberg-McQueen, C. M., 103
Spread spectrum, 198
Sprint, 23
Sprint PCS, 87
   phones, 73
   Wireless Web, 19–20
SSL. *See* Secure Sockets Layer.
St. Laurent, Simon, 104
Standardized Generalized Markup Language (SGML), 119
Standby time, 65–71, 198
StarTAC telephone, 67
State management, 115
   features, 43

State model, 139–143, 187
Statements, 58. *See also* Static conformance statement.
Static conformance statement, 186–189
Stetz, Penelope, 104
String parameterization, 115
Stylized text, 50
Subscriber Identity Module (SIM), 198
Sun Microsystems, 17, 64
Sybase, 19
Symbian, 18

T

Tabindex (attribute), 150–151
Table, 50–51
  (element), 165–167
  example, 167–168
TACS. *See* Total Access Communication System.
Tag. *See* Beginning tag; Ending tag.
  names, encoding, 181
  tokens, 180, 182
Talk time, 65–71, 198
TalkAbout, 67
Target URI, 173
Tasks, 51–55, 128–132
  execution failure, 173
  shadowing. *See* Card/deck task shadowing.
TCP/IP. *See* Transmission Control Protocol/Internet Protocol.
Td (element), 167
TDMA. *See* Time Division Multiple Access.
Teardown, 4
  sessions, 7
Technology/development. *See* exo-net; MainFrame; Sky City Hotels.
Template (element), 147–148
Text, 162–168. *See also* Stylized text.
  support, 42
  (text entry control), 157
Text-based protocol, 76
Text-entry area, 159
Third generation (3G), 198

wireless, 191
Third-party independent sites, 78
Time Division Multiple Access (TDMA), 9, 10, 16, 67, 70, 198. *See also* IS-136 TDMA; North American TDMA.
TimePort, 67
Timer
  (element), 160–162
  example, 161–162
Tokens. *See* Attribute; Extension tokens; Global extension tokens; Tag. assignment. *See* Wireless Markup Language.
  table. *See* Wireless Markup Language.
Total Access Communication System (TACS), 198
Touch-sensitive screens, 100
Tr (element), 167
Transaction model. *See* Wireless Application Protocol.
Transmission Control Protocol/Internet Protocol (TCP/IP), 11, 27, 198
  connection, 34
  networks, 86
Transport layer, 7, 198
Transports, 5–6
Trickle charge, 199
Triple mode, 199
Two-way pagers, 12
Two-way paging, 16
type (attribute), 55–56

U

UA, 135
UI. *See* User Interface.
UMTS. *See* Universal Mobile Telecommunications Services.
Uniform Resource Locator (URL), 33, 34, 37, 40, 42, 54, 62, 72, 119. *See also* Author-specified URL; Relative URLs; Resources;

Wireless Markup Language.
  Demo, 53
  entering, 62, 88
  input dialog, 143
  manipulation, 44
  request, 127
  retrieval, 89
  schemes, 120
  viewing, 94
Uniform Resource Name (URN), 119
Universal Mobile Telecommunications Services (UMTS), 199
Unknown DTD, 171
unknown (event), 55
Unwired Planet, 13, 14, 40
UP.Browser. *See* Phone.com.
UP.Link gateway, 77
UP.Simulator. *See* Phone.com.
URI, 128, 129, 145, 169. *See also* Target URI.
  attribute value, 171
  syntax, 130
URL. *See* Uniform Resource Locator.
URN. *See* Uniform Resource Name.
User agent, 126, 148, 158, 165–169. *See also* Wireless Markup Language.
  limited history, 170
  semantics, 170–175, 187
User input, support, 42
User interaction, 44
User Interface (UI), 119. *See also* Navigational user interface.
  element, 133
Username/password pars, 80
User-visible side effects, 131
UTF-8/UTF-16, 120, 185

V

value (attribute), 56, 154, 161
Variable substitution syntax, parsing, 142
Variables, 58, 123, 139–143. *See also* Wireless Markup Language.

Index

encoding, 181
  setting, 142–143
  substitution, 140–141
  validation, 143
vCalendar, 40
vCard, 40
%vdata, validation, 181
Vdata type, 124
Verizon, 20. *See* Mobile Web.
  WAP services, 89
  Wireless, 87
Vocoder, 199
Voice
  activation, 71
  mail, 199
    screening, 3
  messaging, 199
  recognition, 56–57
Voice-activated dialing, 199
Voice/data networks, 11–12
Voice-operated transmitter (VOX), 199
VOX. *See* Voice-operated transmitter.

W

W3C. *See* World Wide Web Consortium.
WAE. *See* Wireless Application Protocol Application Environment.
WAP. *See* Wireless Application Protocol.
WAP gateways, 75–80
WAPIT, Ltd., 18
WAPMan. *See* Wireless Application Protocol.
WAP-Resources.net, 107
WBMP. *See* Wireless Bitmap.
WBXML. *See* Wireless Binary XML.
WCDMA. *See* Wideband CDMA.
WDP. *See* Wireless Datagram Protocol.
Web Clipping. *See* World Wide Web.
Web sites. *See* Wireless Application Protocol.
Web transaction model, 33–35
White Pages, 87

White space, 162, 168
Wideband CDMA (WCDMA), 199
Windows CE, 61
Windows NT Server, 83
WinWAP, 63–64
Wireless Application Protocol Application Environment (WAE), 38, 45–58, 116, 119
  examination, 40–44
Wireless Application Protocol (WAP), 1, 11, 119, 199. *See also* WinWAP.
  application, 92
  environment, 45–58
  architecture, 14, 37–39
  background. *See* exo-net; MainFrame; Sky City Hotels.
  binary encoding, 27
  binary file, 36
  browser, 18, 60, 97. *See also* Ericsson.
  capabilities, 89
  client software, 59
  consumer profile, 86–89
  convergence, 21–25
  devices, 27, 35, 64–72, 88, 91
    characteristics, 25–26
  Forum, 13, 14, 16–17, 22, 106
    members, 17–19
  future, 97
  gateways, 27, 51, 75, 80, 107
    services, 76–78
  hardware, 59
  history, 13
  Hole Sun, 106
  infrastructure creators, 18
  need, 26–27
  next generation, 101
  origins, 13–16
  overview, 29
  portals, 20
  problems, 97–99
    solution, 99–101
  profiles, 81
  protocol, 4, 16
  search engines, 73
  security contribution, 80
  server, 4
  services, 88. *See also* Verizon.

availability, 19–20
  sites, 71, 72
  step-by-step usage, explanation, 36
  strategy, implementation. *See* Enterprise WAP strategy.
  telephone, 98
  terminology, 75–76
  theory, practice contrast, 88–89
  transaction model, 35–36
  vendor, 80
  WAP88.com, 74
  WAPdrive, 73
  WAPMan, 62–63
  web sites, 59. *See also* Consumer WAP sites.
Wireless Application Protocol Wireless Markup Language (WAP WML), 31, 109–190
  cards, 143
  character set, 120–124
  data types, 124–126
  decks, 132–139, 143–168
    structure, 143
  definitions/abbreviations, 117–119
  device types, 119
  document status, 115–116
  do types, 133–134
  global extension tokens, 180
  informative references, 117
  Inter-Card navigation, 171–172
  navigation and handling models, 126–139
  normative references, 116–117
  reserved words, 180
  scope, 115, 139
  syntax, 122–124
  tokens, 180–186
  variables, 139, 140–143, 170, 180, 181
Wireless Binary XML (WBXML), 117, 121, 181, 185
XML Content Format, 180
Wireless Bitmap (WBMP), 51

Wireless communications, future, 12
Wireless data
  concepts, 1–8
  introduction, 1
Wireless Data Forum, 107
Wireless Datagram Protocol (WDP), 39
Wireless Developer Network, 107
Wireless Markup Language (WML), 14, 40–43, 46–48, 57–58, 199. *See also* Wireless Application Protocol Wireless Markup Language.
  anchors, 54
  authors, 134, 143
  binary encoding, 78
  character set, 120–122, 186–187
  code, 42
  compact binary representation, 180–186
  data types. *See* Core WML data types.
  decks, structure, 143–170
  document, 189
  (element), 144
  encoder, 188–189
  encoding, 186–187
    examples, 185–186
  extension token assignment, 181–182
  features, 49
  HTML translation, 77–78
  Media Type, 173, 181
  parser, 142
  reference information, 173–180
  reference-processing model, 121
  specification, 109–190
  syntax, 122–124
  token table, 188
  URLs, 119–120
  user agent, 162, 186–188
  validation, 189
  variables, 180
  WMLBrowser, 58
  WMLScript, 14, 18, 19, 29, 38, 40, 43–44, 199
Wireless protocols, 95
Wireless Session Protocol (WSP), 38, 119
  header, 147
Wireless technologies, 8–11, 23
Wireless Telephony Application Interface (WTAI), 44
Wireless Telephony Application (WTA), 38, 40
Wireless Transaction Protocol (WTP), 39
Wireless Transport Layer Security (WTLS), 39, 80
WML. *See* Wireless Markup Language.
WMLScript. *See* Wireless Markup Language.
World Wide Web Consortium (W3C), 119
World Wide Web (WWW / Web), 82, 199
  browser, 22, 49, 76
  page, 98, 106
  servers, 35
  sites, 103–107. *See also* Consumer WAP sites.
Web Clipping, 22
wow-com, 107
WSP. *See* Wireless Session Protocol.
WTA. *See* Wireless Telephony Application.
WTLS. *See* Wireless Transport Layer Security.
WTP. *See* Wireless Transaction Protocol.

X

X.25, 199
XML. *See* Extensible Markup Language.
xml:lang (attribute), 125

Y

Yahoo!, 20, 73
Yellow Pages, 87
Yes2WAP, 107
YourWAP.com, 107